Achieve

AF767764

KS2 SATS MATHS
QUESTION WORKBOOK
MATHEMATICS
EXPECTED STANDARD

YEAR 6
AGE 10–11

STEPH KING
TREVOR DIXON
SARAH-ANNE FERNANDES

hachette LEARNING

Hachette UK's policy is to use papers that are natural, renewable and recyclable products and made from wood grown in well-managed forests and other controlled sources. The logging and manufacturing processes are expected to conform to the environmental regulations of the country of origin.

To order, please visit www.HachetteLearning.com or contact Customer Service at education@hachette.co.uk / +44 (0)1235 827827.

ISBN: 978 1 03602 279 2

First published in 2015

Second edition published in 2018

This edition published in 2026 by Hachette Learning (a trading division of Hodder & Stoughton Limited),

An Hachette UK Company

Carmelite House

50 Victoria Embankment

London EC4Y 0DZ

www.HachetteLearning.com

The authorised representative in the EEA is Hachette Ireland, 8 Castlecourt Centre, Dublin 15, D15 XTP3, Ireland (email: info@hbgi.ie)

Impression number 10 9 8 7 6 5 4 3 2 1

Year 2030 2029 2028 2027 2026

Authors: Steph King, Trevor Dixon and Sarah-Anne Fernandes

Series Editor: Sarah-Anne Fernandes

Accessibility Reviewer: Vivien Kilburn

Illustrations by Ann Paganuzzi

Typeset in India

Printed and bound in Great Britain by Bell & Bain Ltd, Glasgow

A catalogue record for this title is available from the British Library.

Contents

Welcome to Achieve Mathematics: The Expected Standard – Question Workbook

In this book you will find lots of practice and information to help you achieve the expected standard in the Key Stage 2 Mathematics tests.

About the Key Stage 2 Mathematics Tests

The tests will take place in the summer term in Year 6. They will be done in your school and will be marked by examiners – not by your teacher.

There are three papers to the tests:

Paper 1: Arithmetic – 30 minutes (40 marks)

- These questions assess confidence with a range of mathematical operations.
- Most questions are worth 1 mark. However, 2 marks will be available for long multiplication and long division questions.
- It is important to show your working – this may gain you a mark in questions worth 2 marks, even if you get the answer wrong.

Papers 2 and 3: Reasoning – 40 minutes (35 marks) per paper

- These questions test mathematical fluency, solving mathematical problems and mathematical reasoning.
- Most questions are worth 1 or 2 marks. However, there may be one question worth 3 marks.
- There will be a mixture of question types, including multiple-choice, true/false or yes/no questions, matching questions, short responses such as completing a chart or table or drawing a shape, or longer responses where you need to explain your answer.
- In questions that have a method box, it is important to show your method – this may gain you a mark, even if you get the answer wrong.

You will be allowed to use a pencil/black pen, an eraser, a ruler, an angle measurer/protractor and a mirror. You **are not allowed** to use a calculator in any of the test papers.

Test techniques

Before the tests

- Try to revise little and often, rather than in long sessions.
- Choose a time of day when you are not tired or hungry.
- Choose somewhere quiet so you can focus.
- Revise with a friend. You can encourage and learn from each other.
- Read the 'Top tips' throughout this book to remind you of important points in answering test questions.
- Keep track of your score using the table on the inside back cover of this book.

During the tests

- READ THE QUESTION AND READ IT AGAIN.
- If you find a question difficult to answer, move on; you can always come back to it later.
- Always answer a multiple-choice question. If you really can't work out an answer, try to think of the most sensible response and read the question again.
- Check to see how many marks a question is worth. Have you written enough to 'earn' those marks in your answer?
- Read the question again after you have answered it. Make sure you have given the correct number of answers within a question, e.g. if there are two boxes for two missing numbers.
- If you have any time left at the end, go back to the questions you have missed.

Where to get help:

- **Pages 6–10** practise number and place value.
- **Pages 11–19** practise number – addition, subtraction, multiplication and division.
- **Pages 20–28** practise number – fractions, decimals and percentages.
- **Page 29** practises ratio and proportion.
- **Pages 30–32** practise algebra.
- **Pages 33–41** practise measurement.
- **Pages 42–46** practise geometry – properties of shapes.
- **Pages 47–50** practise geometry – position and direction.
- **Pages 51–56** practise statistics.
- **Pages 57–64** provide the answers to the questions.

The pencil icon appears next to questions for which you should show your workings.

Place value of whole numbers

To achieve the expected standard, you need to:
★ know the **place value** of digits in whole numbers up to 1,000,000 and begin to know the place value of digits in whole numbers up to 10,000,000.

1 Write the value of the digit **9** in each of these numbers.

1,904   29,456   4,125,692   1,932,400

________ ________ ________ ________

(1 mark)

2 Draw lines to match the numbers with the value of the digit shown.

7 thousand	1,233,420
900 thousand	597,302
30 thousand	3,095,799
3 million	979,200

(2 marks)

3 Use **all** of the digit cards to make **two** different numbers that can be placed on the shaded part of this number line.
You can only use each card once.

0 1 2 3 4 5 6 7 8

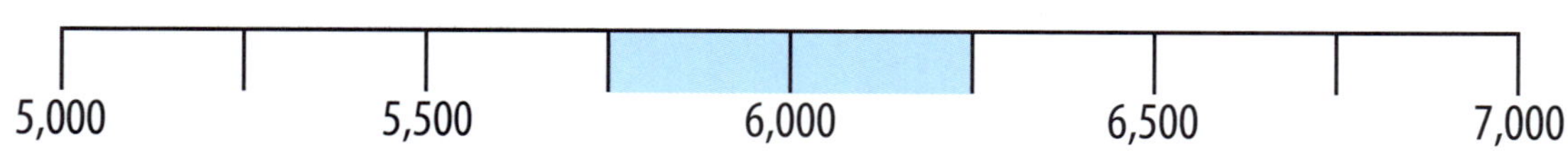

[ ] [ ]

(1 mark)

4 Circle the **two** numbers that have a difference of 20,000

134,200   3,463,400   3,263,400   136,200   3,483,400

(1 mark)

5 Circle **all** the numbers that are less than one million, one hundred and one thousand.

1,150,000   1,005,000   1,050,000   1,105,000

(1 mark)

Top tip
- Find the missing values on a number line first.

/6
Total for this page

Comparing and ordering whole numbers

To achieve the expected standard, you need to:
- ★ use place value of digits in whole numbers to compare and order numbers up to 1,000,000
- ★ begin to work with numbers up to 10,000,000.

1 Write in the missing sign <, > or = to make each statement true.

a) 34,601 ☐ 34,599   b) 709,898 ☐ 709,988

☐ 1 *(1 mark)*

2 Write these masses in order from **heaviest** to **lightest**.

94,500 g   940,500 g   904,500 g   9,500 g   95,400 g

heaviest … **lightest**

☐ 2 *(1 mark)*

3 Gus puts these five numbers in their correct places on a number line.

4,109   5,299   5,881   4,090   4,801

a) Write the number closest to 5,000 ☐

☐ 3a *(1 mark)*

b) Write the number furthest from 5,000 ☐

☐ 3b *(1 mark)*

4 This table shows the total number of visitors to the zoo each year.

2012	2013	2014	2015	2016	2017
695,099	695,100	607,313	700,995	670,133	617,113

Write the years in the correct order in the table below. One has been done for you.

	Year
Greatest number of visitors	2015
Fewest number of visitors	

☐ 4 *(1 mark)*

Top tip

- When comparing numbers, write the numbers in place value columns and make sure the columns line up.

☐ /5 Total for this page

Rounding

To achieve the expected standard, you need to:

★ **round** any whole number to the nearest 10; 100; 1,000; 10,000; or 100,000.

1 Write digits in the boxes to make this statement true.

☐78 and 5☐0 both round to 500 when rounded to the nearest hundred.

1 (1 mark)

2 Round these lengths to the **nearest** 10 cm.

a) 125 cm ☐ cm

b) 99 cm ☐ cm

c) 543 cm ☐ cm

2 (1 mark)

3 Circle the numbers that round to 1,000 when rounded to the nearest **hundred**.

949 1,050 952 1,045 1,100 1,009

3 (1 mark)

4 Round each of these numbers as described.

	To the nearest 10	To the nearest 1,000	To the nearest 10,000
45,419			
234,549			
3,126,095			

4 (2 marks)

5 Ashton thinks of a number and rounds it as shown in the table below.

	To the nearest 1,000	To the nearest 100	To the nearest 10
Ashton's number	6,000	5,700	5,750

Find the smallest number and the largest number that Ashton could have used.

a) Smallest: ☐

b) Largest: ☐

5a 5b (2 marks)

/7 *Total for this page*

Place value of decimal numbers

To achieve the expected standard, you need to:
★ know the place value of digits in numbers with up to two decimal places.

1 3 ÷ [ ] = 0.3

1 (1 mark)

2 Fill in the missing numbers.

9.3 8.8 [ ] 7.8 [ ] [ ]

2 (1 mark)

3 Pete counts on five steps of **1.5** from 2.5

Write each of the numbers he lands on.

2.5 [ ] [ ] [ ] [ ] [ ]

3 (1 mark)

4 Jennie writes down a number.

It has 4 hundredths, no tenths and 8 ones.

What is Jennie's number? [ ]

4 (1 mark)

5 Here is a set of decimal numbers:

0.6 6.66 6 0.06

Ben says 6 is the smallest number.

Explain why Ben is incorrect.

____

____

5 (1 mark)

6 Write these numbers in order of size, starting with the smallest.

7.2 20.7 70.02 2.7 7.02

[ ]	[ ]	[ ]	[ ]	[ ]
smallest				**largest**

6 (1 mark)

Top tip

- When comparing numbers, write the numbers in place value columns and make sure the columns line up.

/6 Total for this page

Negative numbers

To achieve the expected standard, you need to:
- ★ use negative numbers in practical contexts
- ★ calculate intervals across zero.

1 Fill in the missing numbers in this sequence.

7 4 ☐ –2 ☐

1 (1 mark)

2 Sam parks his car on **Level –2** of the town car park.

How many levels must he go up to reach the exit to the shops on **Level 3**? ☐

3
2
1
0
–1
–2

2 (1 mark)

3 The thermometer shows the temperature recorded in France on a day in June.

The temperature recorded on a day in December was 32°C **lower**.

What was the temperature in December? ☐ °C

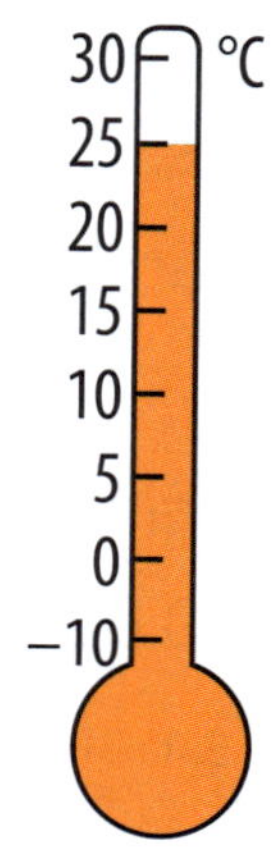

3 (1 mark)

4 Paulo picks two number cards. One of the numbers is negative and one of the numbers is positive.

The difference between the two numbers is 11

a) Find **two** different pairs of numbers he could pick.

☐ and ☐

☐ and ☐

4a (1 mark)

b) Jan said that one of the numbers could be 11

Is Jan correct? Circle your answer. YES / NO

Explain your answer.

____

____

____

4b (1 mark)

/5 Total for this page

Addition

To achieve the expected standard, you need to:

- ★ secure adding whole numbers up to four digits
- ★ add whole numbers with more than four digits
- ★ use formal written methods.

1 Write the missing digits. 9,000 + 7,000 + □ = 16,800

1 (1 mark)

2 72,545 + 48,938 =

2 (1 mark)

3 a) How many points do the Red and Green teams score **altogether**?

3a (1 mark)

b) What is the **total** number of points scored by all three teams?

3b (1 mark)

4 Write the missing digits to make the addition correct.

	□	9	4	8
+	2	□	4	5
	6	0	□	□

4 (1 mark)

5 A marathon run raised £235,748 for a charity in 2016

In 2017 the run raised £79,200 more than in 2016

What is the **total** amount of money raised so far? £

5 (2 marks)

Top tip

- Write down any numbers that you need to carry in the correct column so you remember to include them.

/7 Total for this page

Subtraction

To achieve the expected standard, you need to:

★ secure subtracting whole numbers up to four digits
★ subtract whole numbers with more than four digits
★ use formal written methods.

1 34,562 – 23,638 =

1 (1 mark)

2 Find the difference between 450,000 and 290,000 □

2 (1 mark)

3 How many bags of carrots were sold?

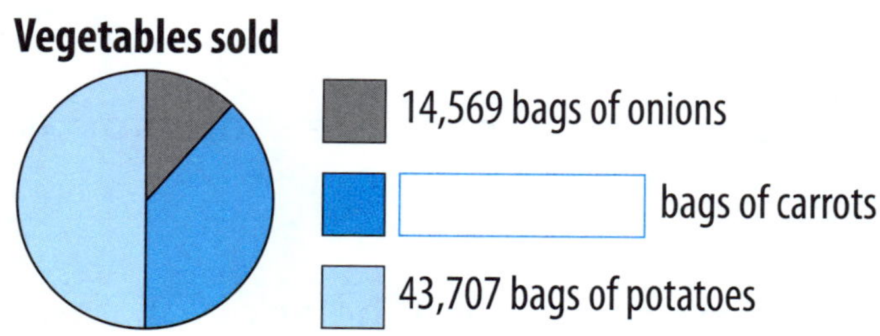

3 (1 mark)

4 The hospital is collecting money for charity. Their target is to collect £15,550

The doctors have donated £2,759. The patients have donated £5,764. The coffee shop in the hospital has raised £1,905

How much **more** money do they need to reach their target?

□

4 (2 marks)

5 Write the missing digits to make the subtraction correct.

	4	0	□	8	6
–	2	□	6	3	□
	1	6	2	□	9

5 (1 mark)

Top tip

- Remember to check your answer using the inverse operation: addition.

/6 Total for this page

Multiplying and dividing by 10 and 100

To achieve the expected standard, you need to:

★ multiply and divide whole numbers and decimal numbers with up to two decimal places by 10 or 100.

1 Fill in the missing numbers.

a) 235 × 100 = ☐ b) 745 ÷ ☐ = 7.45

c) ☐ × 100 = 99

☐ 1 (1 mark)

2 Draw lines to match each calculation with the correct answer.

456 × 100	45
45.6 ÷ 10	0.456
0.45 × 100	4.56
4.56 ÷ 10	45,600

☐ 2 (1 mark)

3 Asha draws a plan of her new sandpit.

A length of 100 cm on the real sandpit is drawn as 10 cm on her plan.

Fill in the missing dimensions on Asha's plan.

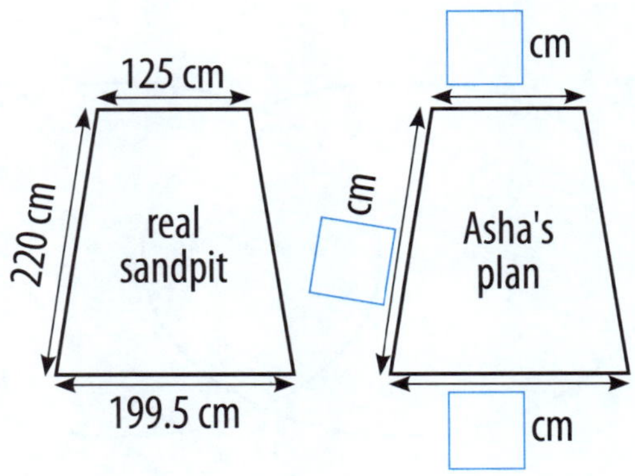

☐ 3 (2 marks)

4 Use the clues to find the mass of the other parcels.

- Parcel C is **100 times lighter** than parcel A.
- Parcel A is **10 times heavier** than parcel B.

Parcel A is ☐ kg and parcel B is ☐ kg.

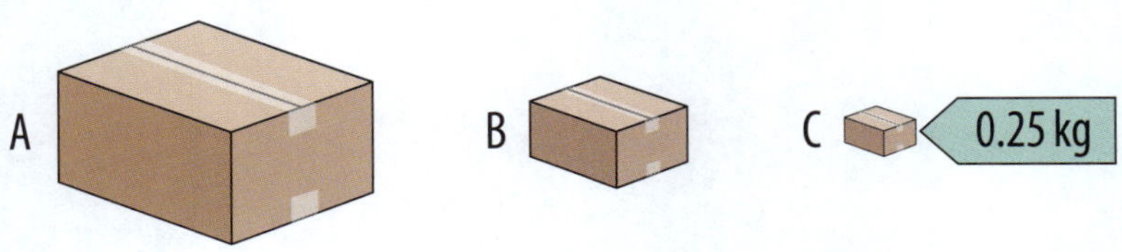

☐ 4 (1 mark)

Top tip

- Read the question carefully and decide which operation you need to use.

☐ /5 Total for this page

Multiples and factors

To achieve the expected standard, you need to:

★ recognise and use **multiples** and **factors**.

1 Circle **all** the multiples of **11** listed below.

77 220 144 177 1,100 111

1 *(1 mark)*

2 Sophie is thinking of an **odd** number.

It is a factor of 24 **and** a factor of 15

What is the number? ☐

2 *(1 mark)*

3 Here are five numbers.

2 3 4 5 6

3 *(2 marks)*

Write each number once into the Venn diagram.

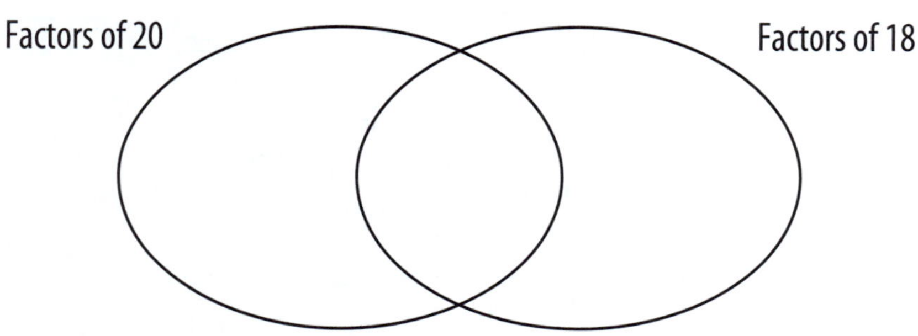

4 Complete the wheel for the number 32 by writing all the factors of 32 in the spaces on the wheel.

4 *(1 mark)*

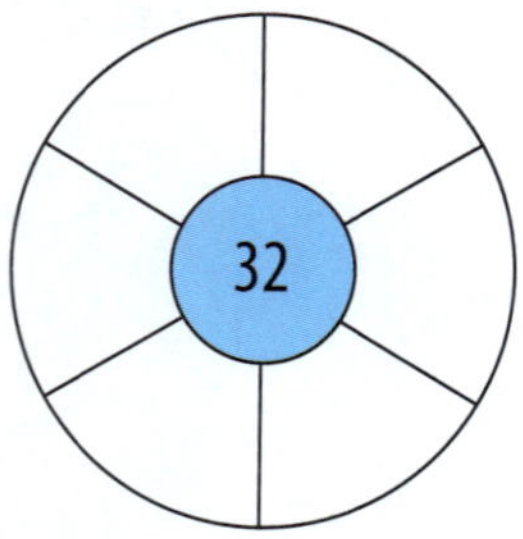

5 Write **three** factors of 30 that are not factors of 15

☐ ☐ ☐

5 *(1 mark)*

/6 *Total for this page*

Multiplying by larger numbers

To achieve the expected standard, you need to:

★ draw upon multiplication facts up to 12 × 12 and place value to:
- multiply numbers with up to four digits by a one-digit number using short multiplication
- multiply numbers with up to four digits by a two-digit number using the formal long multiplication method and become more confident with larger numbers.

1 3,609 × 7 =

1 (1 mark)

2

4	3	4	5
×		2	3

2 (2 marks)

3 Stickers come in boxes of 525

A shop orders 17 boxes.

How many stickers are there altogether in 17 boxes?

[ ] stickers

3 (1 mark)

4 a)

	2	5	
×		2	6
1		3	6
5	1	2	
	6	5	6

4a (1 mark)

b)

	2		8
×			5
1	0	4	0
	2	4	0
7	2		0

4b (1 mark)

5 A shop orders 52 boxes of oranges.

Each box contains 24 bags of oranges. Each bag contains 12 oranges.

How many oranges does the shop order in total?

[ ] oranges

5 (2 marks)

Top tip

- Always show your method because you could gain a mark even if your answer is incorrect.

/8

Total for this page

Square numbers

To achieve the expected standard, you need to:
★ recognise and use **square numbers** up to 144.

1 Use two different **square numbers** to make this statement true.

☐ – 9 = ☐

1 (1 mark)

2 Jack has **80** counters. He makes different squares using the counters.

This is Jack's square using four counters.

How many counters will Jack use to make his largest square? ☐

2 (1 mark)

3 Find **two** different ways to make this calculation true.

square number		square number		prime number
☐	+	☐	=	☐
☐	+	☐	=	☐

3 (1 mark)

4 Paul says that this square has sides 9 cm long, because 9 × 4 is 36

Amy says the sides are 6 cm long.

Who is correct? ☐

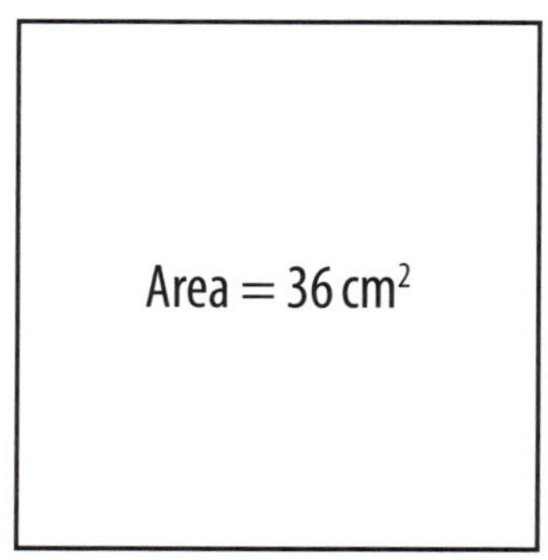

Not to scale

4 (1 mark)

5 Ben says there are two square numbers between 80 and 110

Do you agree? YES / NO

Explain your thinking.

____

____

5 (1 mark)

/5 Total for this page

Short division

To achieve the expected standard, you need to:

- ★ use the formal written method of short division to divide numbers with up to four digits by a one-digit number
- ★ divide whole numbers mentally, drawing upon multiplication facts and place value, and begin to use these facts to work with larger numbers.

1 $434 \div 7 =$

(1 mark) 1

2 $5{,}742 \div 9 =$

(1 mark) 2

3

5,304					
?					

The value of the whole bar is 5,304

Write the value of the shaded part.

(1 mark) 3

4 Each table has 8 seats.

How many tables are needed to seat 349 people?

tables

(1 mark) 4

5 The total mass of 6 different parcels is 2,865 grams.

What is the **mean** mass of the parcels? g

(1 mark) 5

Top tips

- In a problem like question 4, remember that one table may not be completely full.
- You must show a remainder as a fraction or a decimal when calculating the mean as an average.

/5

Total for this page

Long division

To achieve the expected standard, you need to:

★ become more confident using long division to divide numbers with up to four digits by a two-digit whole number.

1

15	8	5	5

(2 marks)

2

24	1	1	2	8		

(2 marks)

3 A car uses **18 litres** of petrol to travel a distance of **594 kilometres**.

How many kilometres does it travel **per litre** of petrol? ______ km

(1 mark)

4 The Parker family has saved the **same** amount of money **each month** for **3 years**.

At the end of 3 years they have saved £2,340

How much did they save each month?

£ ______

(2 marks)

5 A rosette is made from 32 cm of ribbon.

How many rosettes can be made from a **7 metre** piece of ribbon?

______ rosettes

(2 marks)

/9

Total for this page

Top tip

- Always show your method because you could gain a mark even if your answer is incorrect.

Prime numbers

To achieve the expected standard, you need to:
★ recognise and use **prime numbers** less than 20.

1 Complete the list of prime numbers.

☐ 3 5 ☐ ☐ 13 ☐ 19

1 (1 mark)

2 Which prime number is a factor of both 9 and 18? ☐

2 (1 mark)

3 Izi picks one **odd** prime number and one **even** prime number.

The difference between the two prime numbers is 15

Which two prime numbers does she pick?

☐ and ☐

3 (1 mark)

4 Callum is thinking of a number.

It is a factor of 30 and it is prime. What could it be? ☐

4 (1 mark)

5 **42** children get into groups of an **equal** size.

Circle the number of children that is a prime number.

2 groups of 21 children

3 groups of 14 children

6 groups of 7 children

5 (1 mark)

6 The **sum** of three **prime** numbers is **26**

The three numbers are:

☐ + ☐ + ☐

or ☐ + ☐ + ☐

or ☐ + ☐ + ☐

6 (2 marks)

Top tips

- Check you have chosen numbers that match all the criteria in a question.
- Remember that **1** is **not** a prime number. There is only one **even** prime number (2).

/7

Total for this page

Fractions of amounts

To achieve the expected standard, you need to:
★ calculate simple fractions of whole numbers and quantities.

1 a) $\frac{1}{8}$ of 40 = ☐ b) $\frac{3}{8}$ of 40 = ☐ 1 (1 mark)

2 Megan has $\frac{3}{4}$ of £120 and Ali has $\frac{3}{8}$ of £160
How much **more** money does Megan have? ☐ 2 (1 mark)

3 The **whole** container holds **3.6 litres** of water.
How much water is shown here? ☐ litres 3 (1 mark)

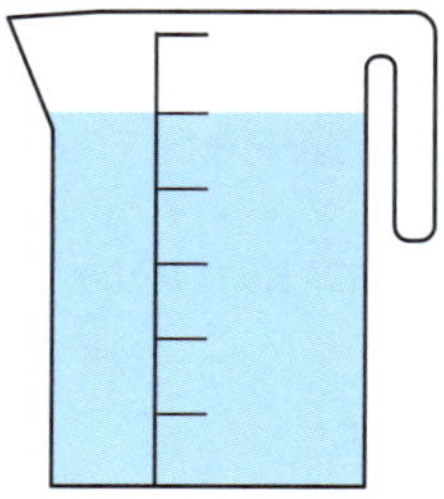

4 Three schools raised £4,800 in total.
a) ☐ was raised by Oak School. 4a (1 mark)
b) ☐ was raised by Ash School. 4b (1 mark)

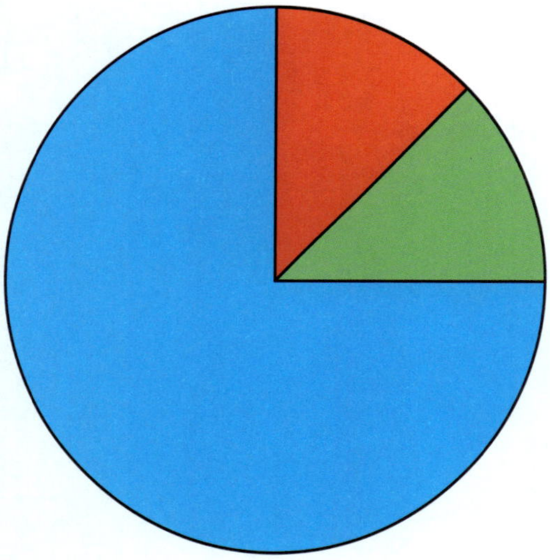

5 £45 is $\frac{\square}{5}$ of £75 5 (1 mark)

6 On Monday, Ben read $\frac{1}{3}$ of his book.
On Tuesday, he read the **other** 72 pages to finish the book.
How many pages are there in Ben's book? ☐ 6 (2 marks)

Top tip
- To find fractions of amounts, always divide by the denominator and then multiply by the numerator.

/8 Total for this page

Mixed numbers

To achieve the expected standard, you need to:

★ recognise **mixed numbers** and **improper fractions** and convert from one to the other.

1 What is $6\frac{2}{3}$ as an **improper fraction**? ☐

(1 mark) 1

2 Rewrite this sequence of improper fractions as **mixed numbers**.

$\frac{5}{4}$ $\frac{7}{4}$ $\frac{9}{4}$ $\frac{11}{4}$ $\frac{13}{4}$

☐ ☐ ☐ ☐ ☐

(1 mark) 2

3 Write the answer to this calculation as a **mixed number**.

$\frac{7}{8} + \frac{5}{8} =$ ☐

(1 mark) 3

4 Complete the empty boxes on this number line.

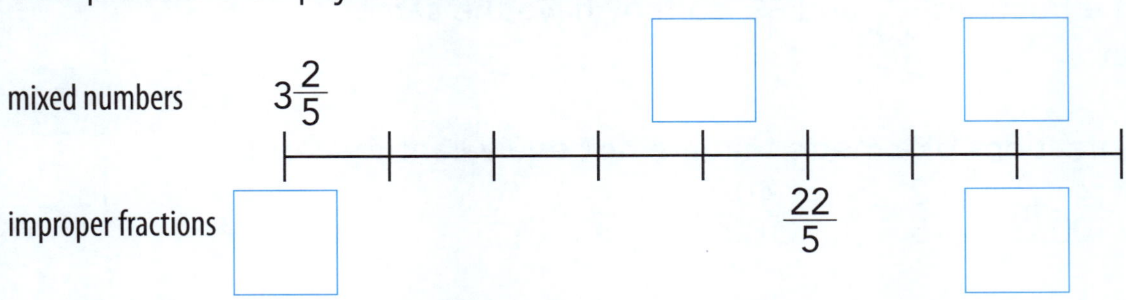

(2 marks) 4

5 Circle **all** the improper fractions that are **equivalent** to **6**

$\frac{12}{6}$ $\frac{30}{5}$ $\frac{18}{3}$ $\frac{60}{12}$ $\frac{54}{9}$

(1 mark) 5

6 Write the amount shaded as a **mixed number** and as an **improper fraction**.

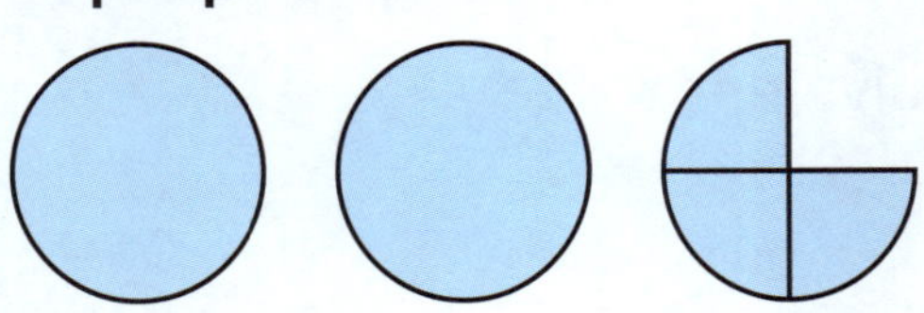

= ☐ ☐

(1 mark) 6

Top tip

- Remember that $\frac{3}{3}$, $\frac{4}{4}$ and $\frac{6}{6}$ are all equivalent to 1.

/7

Total for this page

Equivalent fractions

To achieve the expected standard, you need to:
- ★ recognise and use **equivalent fractions**
- ★ find equivalent fractions with lower denominators
- ★ rewrite a pair of fractions so they share the same denominator.

1 Draw lines to join the equivalent fractions.

$\frac{2}{3}$	$\frac{75}{100}$
$\frac{3}{4}$	$\frac{3}{18}$
$\frac{2}{5}$	$\frac{8}{12}$
$\frac{1}{6}$	$\frac{40}{100}$

1 (1 mark)

2 $\frac{3}{5} = \frac{15}{\square}$

2 (1 mark)

3 Simplify the fractions $\frac{5}{20}$ and $\frac{9}{12}$ so they have the **same** denominator.

3 (1 mark)

4 Circle the fractions of a metre that are **not** equivalent to 40 cm.

$\frac{4}{10}$ m $\frac{4}{100}$ m $\frac{2}{5}$ m $\frac{40}{100}$ m $\frac{1}{4}$ m

4 (1 mark)

5 Rewrite each of these fractions so that they all have the **same** denominator.

$\frac{1}{2}\ \frac{\square}{\square}$ $\frac{2}{5}\ \frac{\square}{\square}$ $\frac{3}{4}\ \frac{\square}{\square}$

5 (1 mark)

6 Shade the diagram to show a fraction that is equivalent to $\frac{15}{24}$

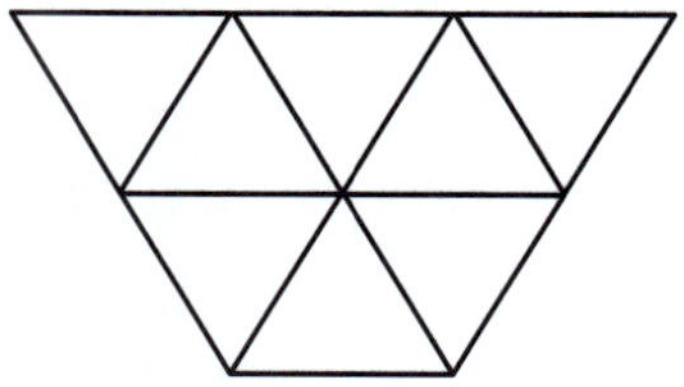

6 (1 mark)

Top tip
- Use your multiplication facts to help find equivalent fractions.

/6
Total for this page

Adding and subtracting fractions

To achieve the expected standard, you need to:
- ★ add and subtract fractions with the same denominator, using mixed numbers where appropriate for the context
- ★ add and subtract fractions with denominators that are **multiples** of the same number, and become more confident with more complex calculations.

1 $1\frac{4}{5} - \frac{2}{5} = \square$ 1 (1 mark)

2 $1\frac{3}{5} - \frac{4}{5} = \square$ 2 (1 mark)

3 $\square = \frac{2}{3} + \frac{5}{6}$ 3 (1 mark)

4 $\frac{3}{4} + \square + \frac{1}{2} = 2$ 4 (1 mark)

5 Draw lines to match these calculations with their answers. 5 (2 marks)

$1\frac{5}{8} - \frac{3}{4}$	$3\frac{1}{2}$
$1\frac{3}{4} - 1\frac{1}{12}$	3
$1\frac{3}{5} + \frac{7}{10} + \frac{7}{10}$	$\frac{7}{8}$
$2\frac{5}{6} + \frac{2}{3}$	$\frac{2}{3}$

6 There is $\frac{3}{4}$ litre of water in a jug. 6 (1 mark)

Jack pours in a further $\frac{3}{8}$ litre of water.

How much water is in the jug altogether?

Write your answer as a mixed number. $\square$ litres

/7 Total for this page

Fractions and their decimal equivalents

To achieve the expected standard, you need to:

- ★ recognise and write decimal equivalents for $\frac{1}{4}$, $\frac{1}{2}$, $\frac{3}{4}$ and any number of fifths, tenths or hundredths and read and write decimals as fractions
- ★ recognise and use thousandths and relate them to tenths, hundredths and decimal equivalents.

1 Fill in the empty boxes on this number line.

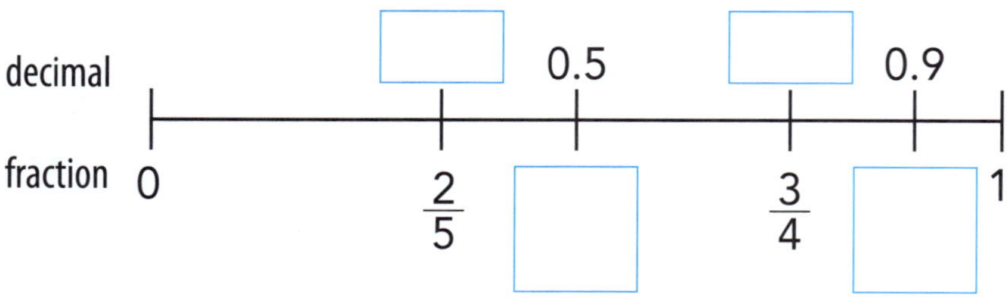

1 (2 marks)

2 $0.36 = \frac{36}{\square}$

2 (1 mark)

3 Circle **all** the fractions that are equivalent to **0.25**

$\frac{25}{100}$ $\quad \frac{2}{5}$ $\quad \frac{5}{20}$ $\quad \frac{1}{4}$ $\quad \frac{4}{12}$

3 (1 mark)

4 Draw lines to match the fraction of the shapes shaded with an equivalent decimal.

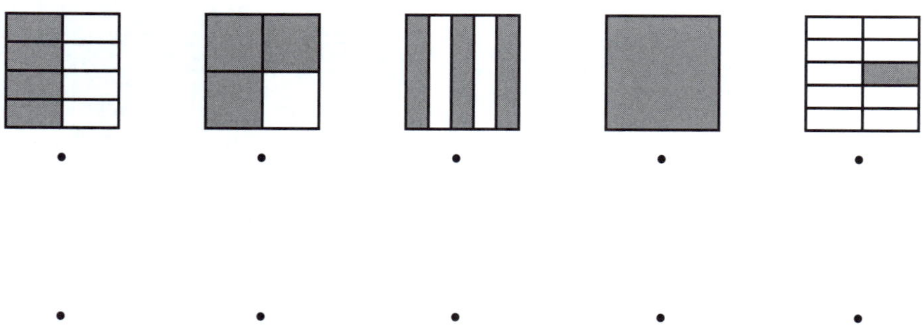

1.0 0.5 0.75 0.1 0.6

4 (1 mark)

Top tip

- Remember, fractions can be converted into decimals by using division (e.g. $\frac{7}{10}$ can be converted using 7 ÷ 10).

/5

Total for this page

Adding and subtracting decimals

To achieve the expected standard, you need to:
★ add and subtract decimal numbers that have the same number of decimal places.

1 a) 64.29 – 20.14 = ☐

b) 7 = 10.9 – ☐

(1 mark) 1

2 Lexi buys these two games.

a) How much does she spend altogether? ☐

b) Work out how much change she will get from £40 ☐

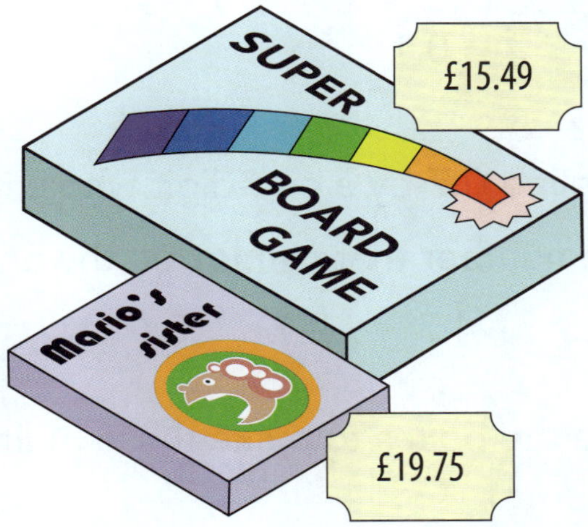

(1 mark) 2a

(1 mark) 2b

3 Find **two** different pairs of numbers that total **6**

1.7 2.46 3.64 4.3 3.54 5.3

☐ and ☐ total 6

☐ and ☐ total 6

(1 mark) 3

4 Ollie jumps 1.26 m. Jahla jumps 0.85 m further.

How far does Jahla jump? ☐ m

(1 mark) 4

5

Container A	Container B	Container C	Container D
2.75 litres	4 litres	2.79 litres	8.32 litres

The **total** water in **two** of these containers is 1.53 litres **less than** the amount in **Container D**.

The **two** containers are ☐ and ☐.

(1 mark) 5

Top tip

- Line up the decimal points when adding and subtracting.

/6 Total for this page

Multiplying decimals

To achieve the expected standard, you need to:
★ multiply a one-digit decimal number by a one-digit number.

1 ☐ × 9 = 6.3 — 1 (1 mark)

2 ☐ = 0.8 × 8 — 2 (1 mark)

3 0.05 × ☐ = 0.1 — 3 (1 mark)

4 Kai multiplied **0.7** by a one-digit whole number.

Circle the number that **cannot** be the answer to his calculation. — 4 (1 mark)

1.4   3.5   4.7   2.1   5.6

5 Complete the boxes to make these multiplications correct. — 5 (2 marks)

0.6 × 3 = ☐

0.6 × 9 = ☐

0.6 × 7 = ☐

0.6 × ☐ = 4.8

6 ☐ kg × 5 = 24 kg ÷ 6 — 6 (1 mark)

7 Calculate the area of this rectangle. ☐ m^2 — 7 (1 mark)

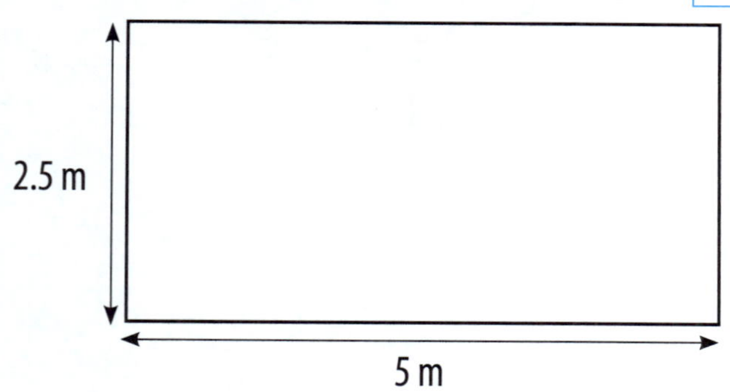

Top tip

- Remember to use multiplication facts and place value to help.

/8 Total for this page

Percentages as fractions and decimals

To achieve the expected standard, you need to:

★ recognise and use the equivalences between simple fractions, decimals and **percentages**

★ become more confident with calculating other decimal fraction equivalents.

1 Complete this table of equivalent fractions, decimals and percentages.

Fraction	$\frac{3}{10}$			$\frac{4}{5}$
Decimal		0.65		
Percentage			5%	

1 (2 marks)

2 Circle the amounts that are **not** equivalent to $\frac{1}{4}$

1.4 25% $\frac{3}{12}$ 0.25 14% 2.5

2 (1 mark)

3 The shaded part of this shape can be described as:

$\frac{\square}{\square}$ or ☐ % or ☐ . ☐

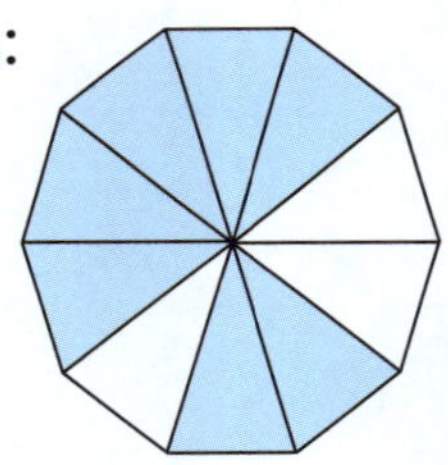

3 (1 mark)

4 Fill in the empty boxes on this number line.

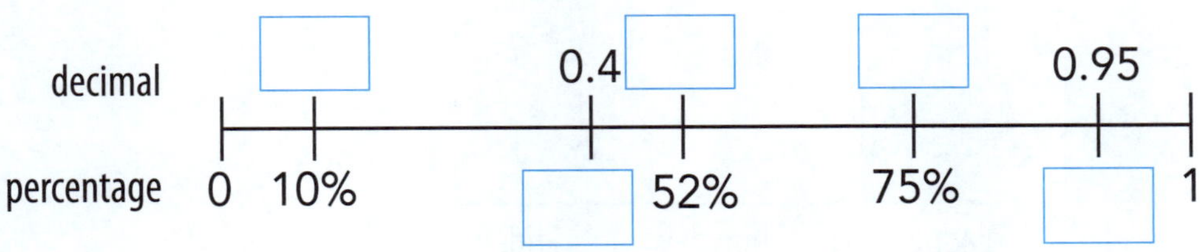

4 (2 marks)

5 George and Alice are sharing a pizza.

George eats **25%** of the pizza.

Alice eats $\frac{2}{5}$ of the pizza.

George thinks that he has eaten the **same** amount of pizza as Alice.

a) Explain why George is incorrect. ______________________________

____

5a (1 mark)

b) Write Alice's fraction of the pizza as a **decimal**. ☐

5b (1 mark)

/8

Total for this page

Finding percentages

To achieve the expected standard, you need to:
★ find simple percentages of whole numbers and quantities.

1 a) 20% of 120 = ☐   b) 25% of 200 = ☐

2 a) 50% of ☐ = 100   b) 25% of ☐ = 100

3 a) 25% of ☐ = £10   b) 75% of ☐ = 15 kg

4 Use the symbol <, > or = to make this statement true.

50% of 300 ☐ 75% of 200

5 This chart shows the number of visitors to a park.

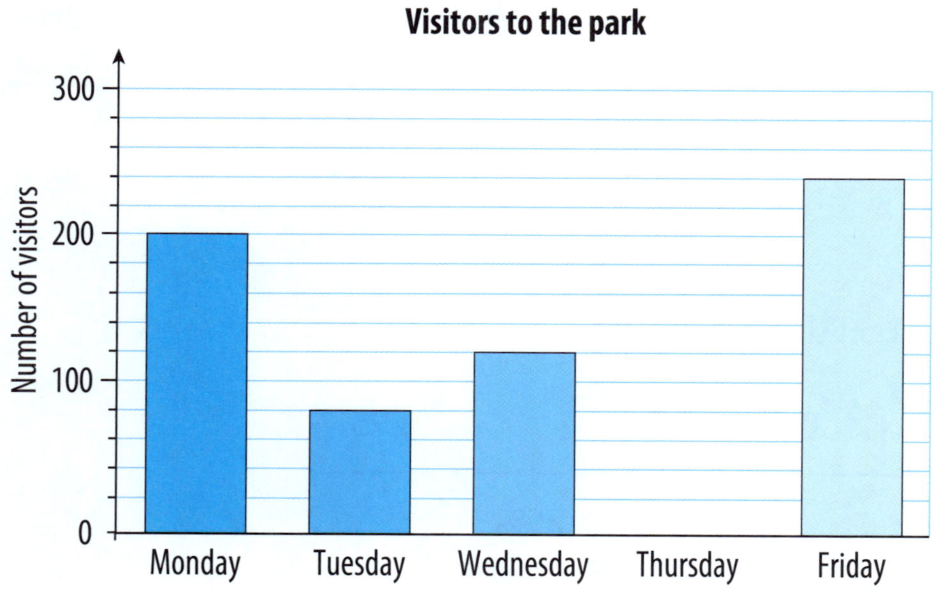

a) The number of visitors on Thursday is 40% of the number of visitors on Monday.

Draw the bar for Thursday on the chart.

b) 30% of the visitors on Friday were children.

How many adults visited the park on Friday? ☐ adults

Top tip

- Finding 20% does not mean divide by 20.
- Check the scale of any bar chart first.

1a (1 mark)
1b (1 mark)
2a (1 mark)
2b (1 mark)
3a (1 mark)
3b (1 mark)
4 (1 mark)
5a (2 marks)
5b (1 mark)

/10 Total for this page

Ratio and proportion

To achieve the expected standard, you need to:
- ★ use simple **ratio** to compare quantities
- ★ estimate the distance on a map using a simple scale.

1 On a shelf there are 3 fiction books to every 2 reference books.

There are 18 reference books on the shelf.

How many fiction books are there? [ ]

1 (1 mark)

2 A recipe uses 100 ml milk for every 250 g flour.

How much milk is needed for **1 kg** of flour? [ ] ml

2 (1 mark)

3 What is the real distance between Ashley Town and Mason Park?

[ ] km

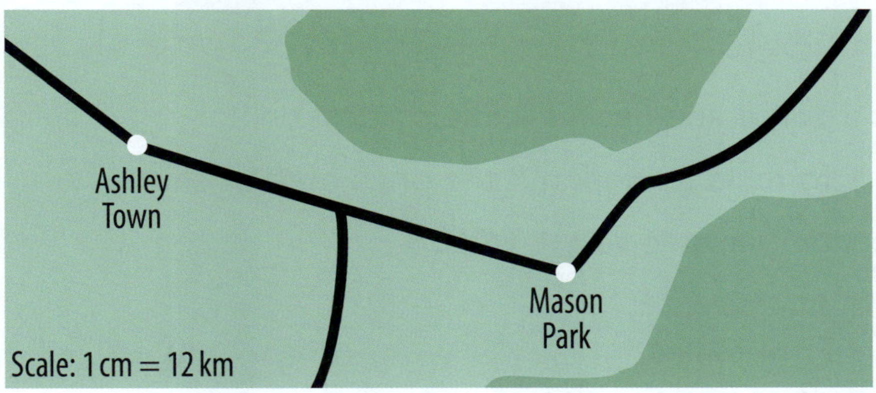

3 (2 marks)

4 In a group of 20 people, 3 people are left-handed.

What **proportion** of the group is **not** left-handed? [ ]

4 (1 mark)

5 A total of 160 points was scored in a quiz.

The Red team scored 5 points **for every** 3 points scored by the Blue team.

How many points were scored by each team?

Red team [ ]

Blue team [ ]

5 (2 marks)

Top tips

- Use your knowledge of multiplication facts to solve direct proportion problems.
- Use factors to help simplify ratios.

/7

Total for this page

Algebra

To achieve the expected standard, you need to:
★ use simple formulae expressed in words.

1 Use the rule to find the missing numbers.

Input	Rule	Output
9 →	multiply by 10 and add 3	→
→		→ 53

2 Use the rule below to fill in the empty boxes.

Double a number and subtract 12

Start number		Answer
24	→	
	→	58
12.5	→	

3 Teddy buys a t-shirt and has his name printed on it.

They use this formula to work out the price of the t-shirt:

Price = £2 x number of letters + £6.50

How much did he pay? £

4 Tick the formula to find the size of Angle *A* if it was a missing value.

$180 - (B - C)$

$180 - (B + C)$

$180 + B - C$

$180 + (B - C)$

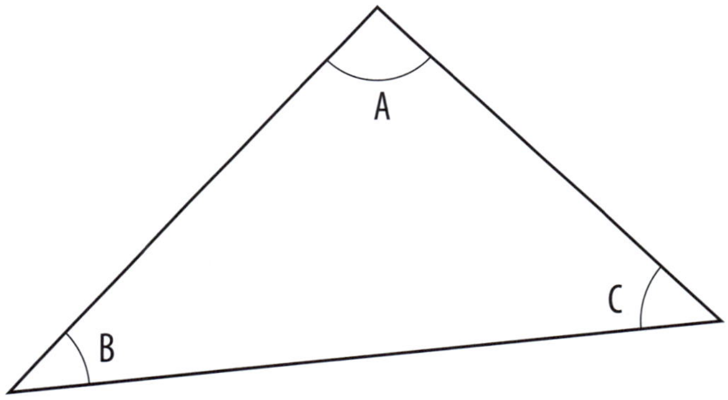

5 a) $n = 8$

What is $4n + 16$?

b) $4q + 10 = 90$

Work out the value of q.

1 (1 mark)
2 (1 mark)
3 (1 mark)
4 (1 mark)
5a (1 mark)
5b (1 mark)
/6 Total for this page

Top tip

- You may want to use brackets to show which part of the calculation to do first.

Sequences

To achieve the expected standard, you need to:

- ★ count forwards or backwards in steps of any whole number with one significant figure
- ★ generate, describe and complete linear number **sequences**.

1 A sequence **starts** at **–15**; the rule is **add 7**

Write the next five terms in the sequence.

–15 ☐ ☐ ☐ ☐ ☐

1 (1 mark)

2 Write the rule for this sequence.

Fill in the missing terms.

The rule is ______________________

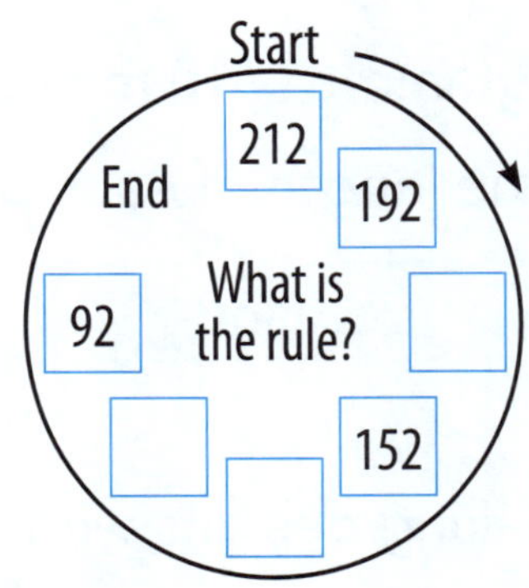

2 (1 mark)

3 Find the rule and the missing terms in this sequence.

148 ☐ 128 ☐ 108

The rule is ______________________

3 (1 mark)

4 The jumps on this number line are of an equal size.

What is the missing value?

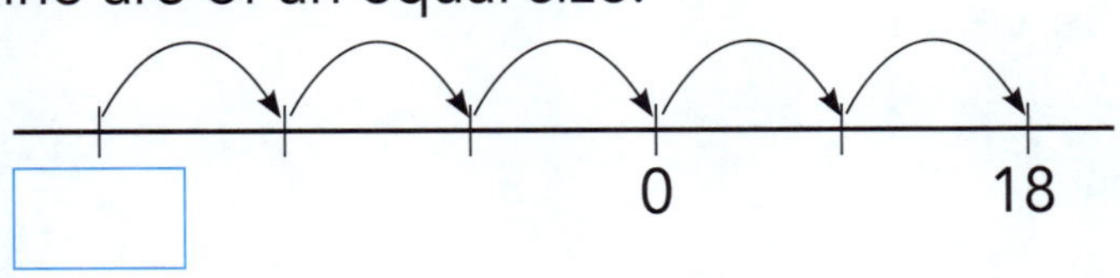

4 (1 mark)

5 A sequence starts at 35

The rule is Add 25

Will the number 725 be in the sequence?

Circle your answer. YES / NO

Explain your answer.

____

____

5 (1 mark)

Top tip

- Look at the number patterns. Think about what stays the same and what changes.

/5 Total for this page

Solving equations

To achieve the expected standard, you need to:

★ find possible values in missing-number problems and equations involving one or two unknowns.

1 Write numbers in the boxes to make this statement true.

$\square + 15 = \square \times 3$

1 (1 mark)

2 Two radio programmes last a total of 60 minutes.

One programme is 10 minutes longer than the other.

What is the length of each programme?

☐ and ☐

2 (1 mark)

3 Mya is making designs with two shapes.

Each shape has a value.

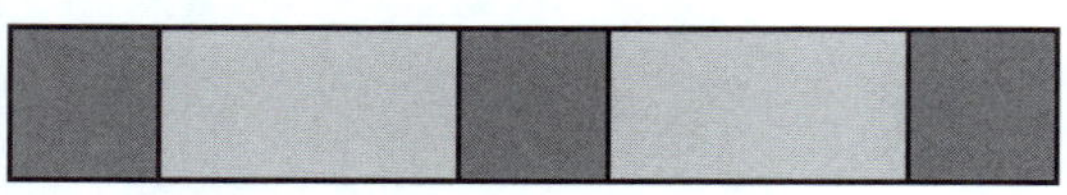 has a value of **85**

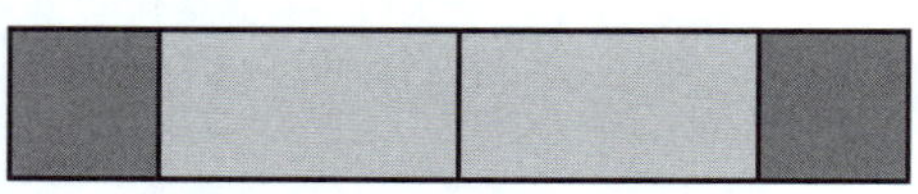 has a value of **70**

What is the value of:

a)  ☐

3a (1 mark)

b)  ☐

3b (1 mark)

4 The pictures show a set of balancing scales and some parcels.

What is the **mass** of each parcel?

Parcel A = ☐ g

Parcel B = ☐ g

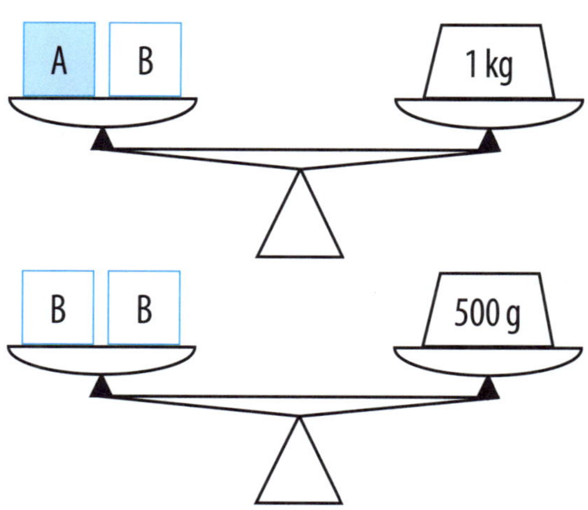

4 (1 mark)

Top tip

- Use inverse operations to help solve missing-number problems.

/5 Total for this page

Length

To achieve the expected standard, you need to:
- ★ use, read, write and convert between metric units of length
- ★ use all four operations to solve problems involving length
- ★ use decimals to two decimal places.

1 2 m + [cm] = 525 cm

(1 mark)

2 How many **25 centimetre** lengths can be cut from a **4.5 metre** rope? [ ]

(1 mark)

3 Draw lines to match the equivalent lengths.

38 cm	38 mm
3.8 cm	380 cm
3.8 m	0.38 m

(1 mark)

4 A car travels a distance of 5.6 kilometres.

How many **metres** is 5.6 kilometres? [m]

(1 mark)

5 What is the **perimeter** of the pitch in metres? [m]

(1 mark)

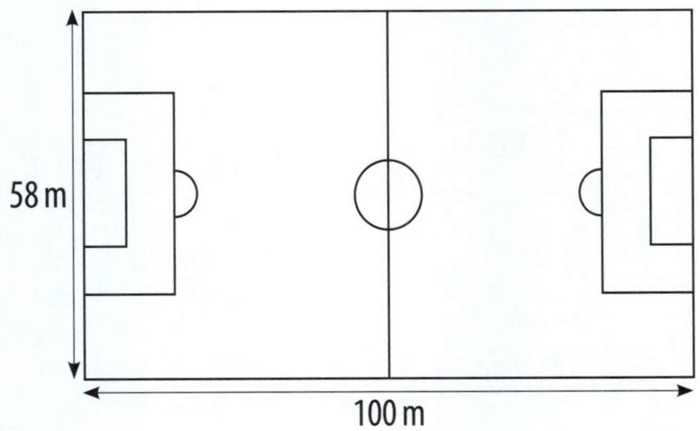

6 The length of a model car is **7.2 centimetres**.

The length of the real car is **60 times** the length of the model.

What is the **length** of the real car?

Give your answer in **metres**. [m]

(2 marks)

Top tip
- Check the unit of measurement in each question and answer.

/7

Total for this page

Mass

To achieve the expected standard, you need to:
- ★ use, read, write and convert between units of **mass**
- ★ use all four operations to solve problems involving mass
- ★ use decimals to two decimal places.

1 What is the total mass of these objects? [g]

1 (1 mark)

2.5 kg   560 g   1.25 kg

2 A recipe uses the mass of potatoes shown on this scale.

What mass of potatoes will be left in a 2.5 kg bag? [ ]

2 (2 marks)

3 All three boxes have the same mass.

What is the mass of **one** of the boxes? [kg]

3 (1 mark)

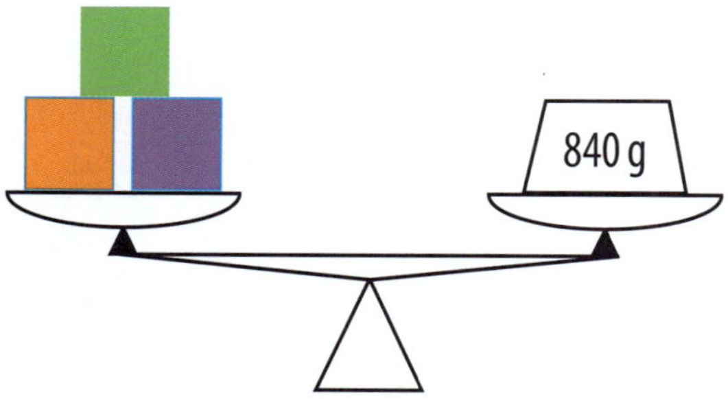

4 A box of 12 tins has a mass of 5.5 kg.

The box without the tins has a mass of 520 g.

What is the mass of each tin? [g]

4 (2 marks)

Top tip
- Writing down the calculations for each step can help organise your thinking, catch errors and may even gain you a mark.

/6
Total for this page

Capacity

To achieve the expected standard, you need to:

- ★ use, read, write and convert between metric units of **capacity**
- ★ use all four operations to solve problems involving capacity
- ★ use decimals to two decimal places.

1 How many millilitres are in 4.65 litres? [ ] ml

(1 mark) 1

2 3 litres = 1.75 litres + [ ] ml

(1 mark) 2

3 How many 250 ml cups can be filled from the water in this jug? [ ]

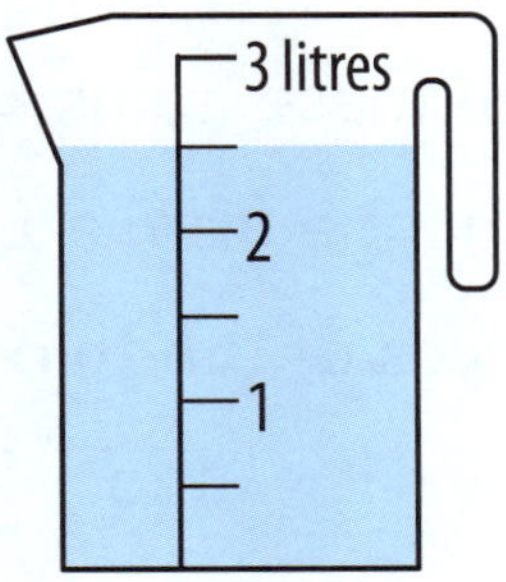

(1 mark) 3

4 A car uses 0.2 litres of petrol to travel a distance of 6 km.
How far will it travel on 1 litre of petrol? [ ] km

(1 mark) 4

5 In a race, teams fill buckets of water that hold 1.25 litres.

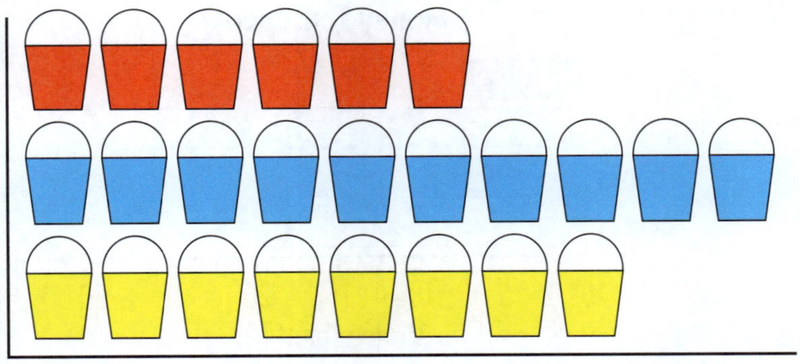

a) Write the total amount of water collected by the Yellow team in **litres**. [ ] l

(1 mark) 5a

b) How much **more** water did the Blue team collect than the Red team? [ ] l

(1 mark) 5b

Top tip

- Check the unit of measurement required each time. If you give an answer in millilitres when the question asks for litres, you will lose marks.

/6
Total for this page

Money

To achieve the expected standard, you need to:

★ use all four operations to solve problems involving money, using decimal places.

1 a) What is the cost of **five** packs of pencils? [ ]

b) What is the cost of **one** ruler? [ ]

1a (1 mark)

1b (1 mark)

2 Sami buys two of these items.

He gets £1.76 change from a £20 note.

What are the prices of the **two** items he bought? [ ] and [ ]

2 (2 marks)

3 Kelsey and Toby have saved £7.65 altogether.

Kelsey has saved £1.25 **more** than Toby.

How much did each child save?

Kelsey [ ] Toby [ ]

3 (2 marks)

4 The bar chart shows how much money Tom spent in different ways last month.

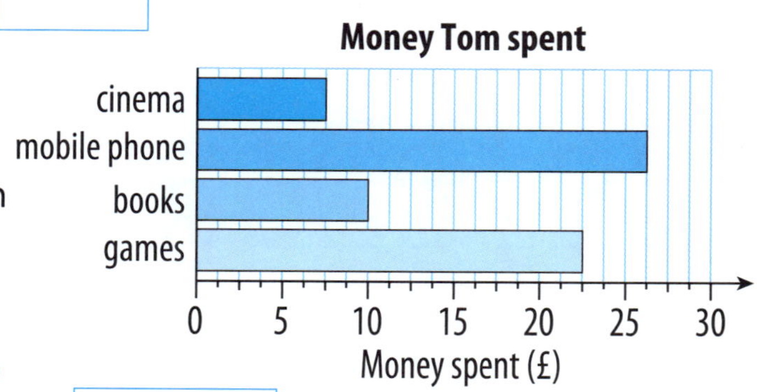

a) How much **more** did he spend on his mobile phone than at the cinema? [ ]

b) How much did Tom spend in **total** last month? [ ]

4a (1 mark)

4b (1 mark)

5 David, Yannis and Marie share £150 prize money.

David's share is $\frac{2}{5}$ of the total prize money and Marie's share is $\frac{1}{3}$ of the total prize money.

How much money does Yannis get? [ ]

5 (2 marks)

Top tip

- Always check you have answered the question that has been asked.

/10 Total for this page

Time

To achieve the expected standard, you need to:
- ★ read, write and convert time between analogue (including clock faces using Roman numerals) and digital 12- and 24-hour clocks
- ★ use a.m. and p.m. where necessary.

1 Draw lines to match these times.

(2 marks)

noon	25 to 8 in the evening
17:35	12 p.m.
19:25	7:25 p.m.
 p.m.	5:35 p.m.

2 Order the following times from **earliest** to **latest**. Write the letters of each clock in the boxes.

(1 mark)

A 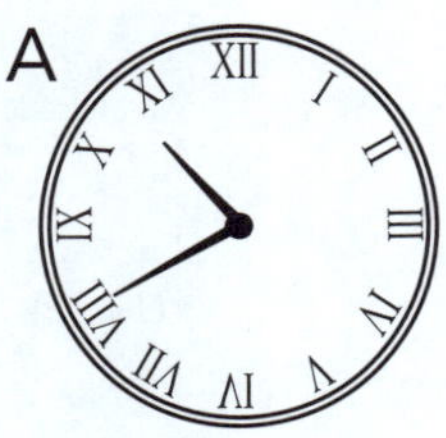 p.m.

B 

C 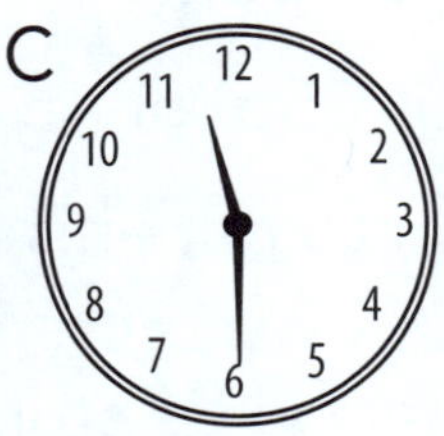 a.m.

D 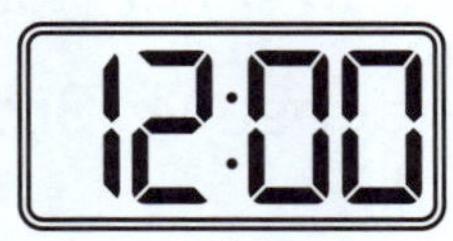

earliest ☐ ☐ ☐ ☐ **latest**

3 Write the time that is 75 minutes **earlier** than the one shown here.

(1 mark)

4 Write the time that is **halfway** between 10:20 a.m. and 13:10

(1 mark)

Top tip
- Think about the number of minutes in an hour.

/5 Total for this page

Time problems

To achieve the expected standard, you need to:

★ calculate the duration of an event using appropriate units of time.

1 A film starts at 9:40 p.m. and ends at 11:25 p.m.

How long does it last? ☐

1 (1 mark)

2 Pete's art lesson ends at 14:45

It lasts for 55 minutes.

What time does it start? ☐

2 (1 mark)

3 The shaded part of the scale shows the length of time that Mr Brown parked his car in the car park.

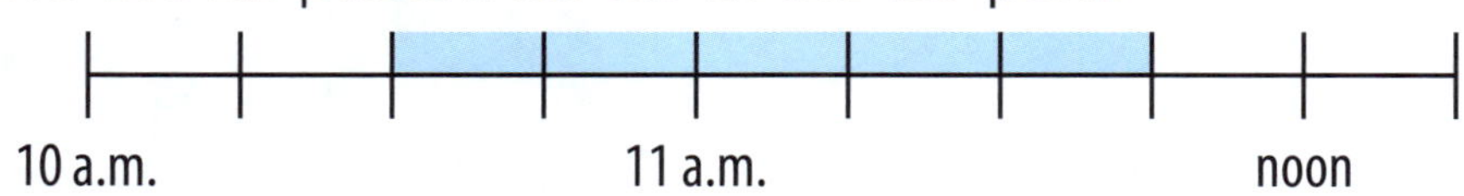

a) What time did he arrive at the car park? ☐

3a (1 mark)

b) The cost of parking a car here is 35p for 15 minutes.

How much did it cost Mr Brown to park his car? ☐

3b (1 mark)

4 This timetable shows the times that buses stop at different places in the town.

Bus garage	10:18	11:18	12:18
Post Office	10:46	11:46	12:46
Station	11:15	12:15	13:15
School	12:08	13:08	14:08

a) How long does it take for the bus to travel from the **bus garage** to the **school**? ☐

4a (1 mark)

b) Mrs Lang arrives at the Post Office bus stop at 11:55

How long must she wait for the next bus? ☐

4b (1 mark)

c) Jasper wants to get the 13:15 bus from the station.

He gets to the station bus stop 18 minutes early.

What time does he get to the station bus stop? ☐

4c (1 mark)

Top tip

- Remember to check any scales first.

/7

Total for this page

Perimeter

To achieve the expected standard, you need to:
★ calculate the **perimeter** of **compound shapes**, when all side lengths are known or can be easily determined.

1 Calculate the perimeter of this square. ☐ cm

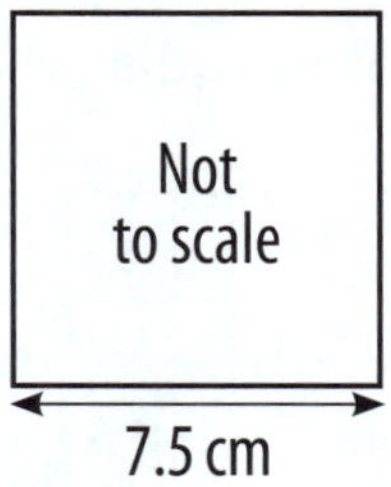

(1 mark) 1

2 This square has a perimeter of 30 cm. What is the length of **one** side? ☐ cm

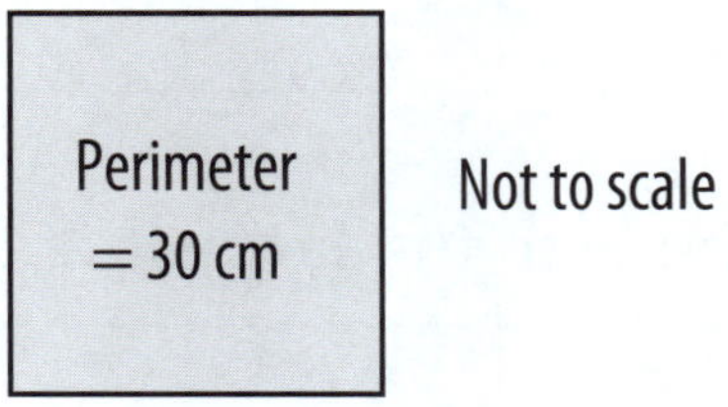

(1 mark) 2

3 This rectangle has a perimeter of 52 cm.

The length of the rectangle is 18 cm. What is the **width** of the rectangle? ☐ cm

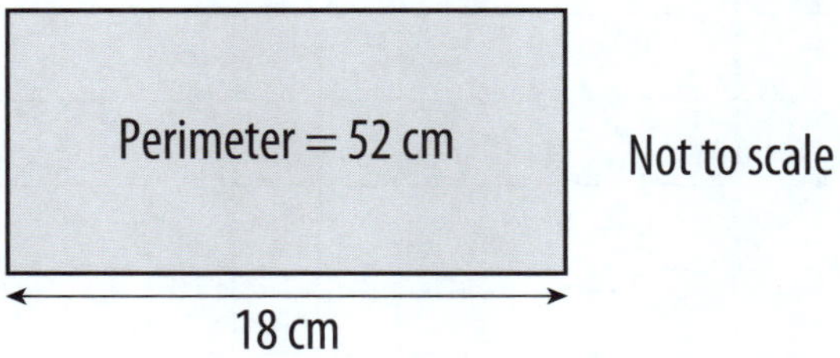

(1 mark) 3

4 A square and a rectangle are placed together to make this new shape.

Calculate the perimeter of the new shape.

Write your answer in **centimetres**.

☐

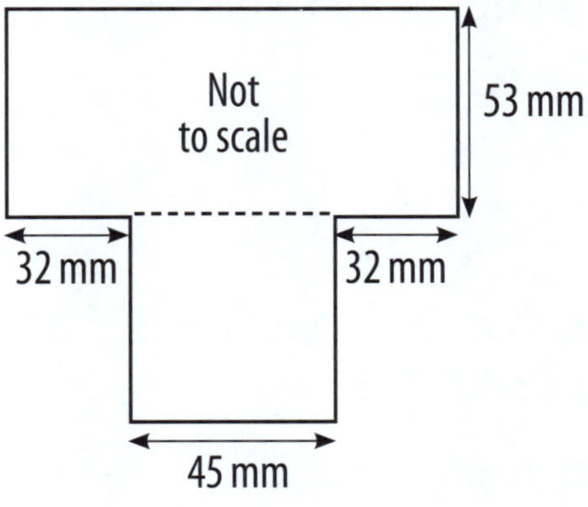

(2 marks) 4

Top tip

- Look for any given measurements. Think how you can use them to work out unknown measurements.

/5 Total for this page

Estimating the area of irregular shapes

To achieve the expected standard, you need to:
★ estimate the **area** of irregular shapes by counting squares.

1 Draw a shape on the grid that has **half** the area of this shape.

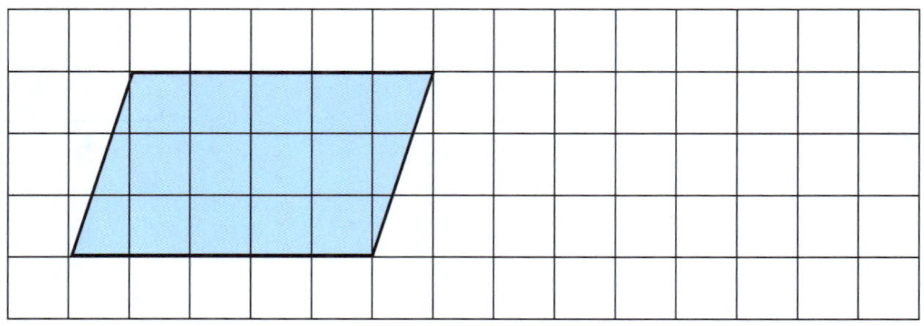

1 *(1 mark)*

2 Tick (✓) the shape with the **largest** area.

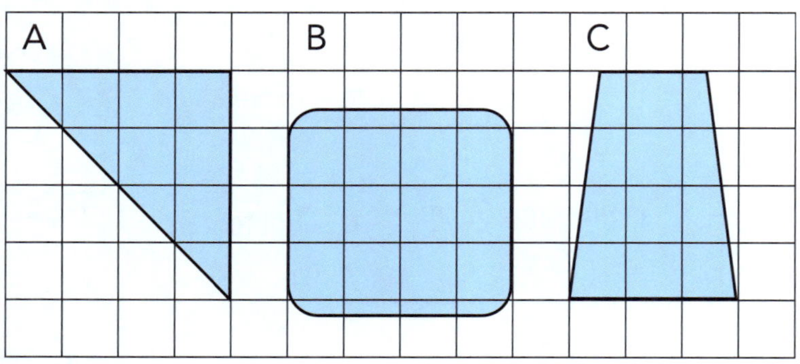

2 *(1 mark)*

3 Estimate the area of this shape.

Each square represents **1 cm²**.

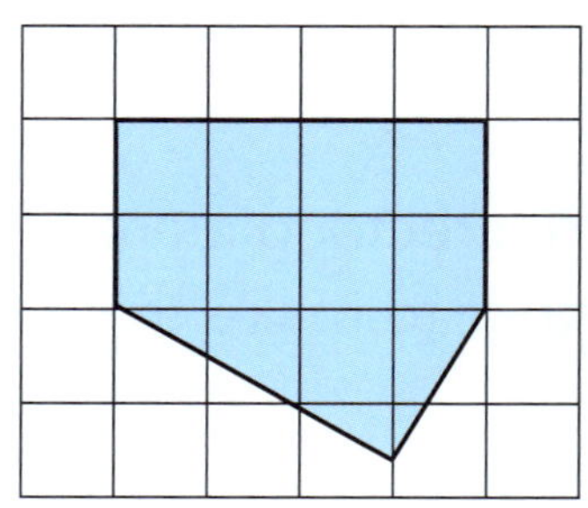

3 *(1 mark)*

4 Asha thinks that the area of this shape is approximately 15 m².

Each square represents **1 m²**.

Do you agree with Asha?
Circle your answer. YES / NO

Explain your answer.

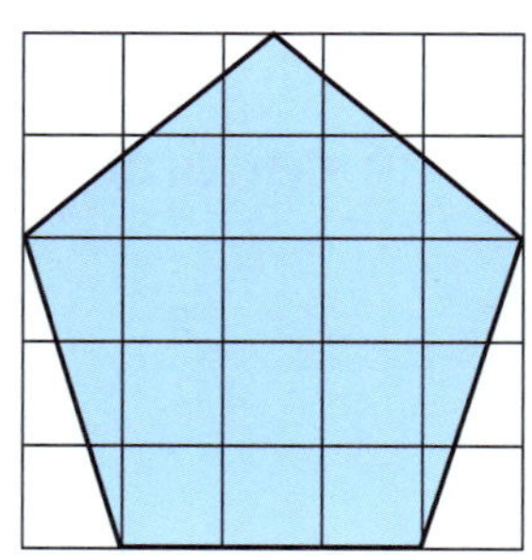

4 *(1 mark)*

Top tip

- Look for fractions of squares that can be put together to make (or nearly make) a whole square.

/4

Total for this page

Area by formula

To achieve the expected standard, you need to:
- ★ calculate and compare the area of squares and other rectangles
- ★ use standard units: square centimetres (cm^2) and square metres (m^2).

1 A square has sides of 11 cm. Calculate its area. ______ cm^2

1 (1 mark)

2 Which of these shapes has the **largest area**? ______

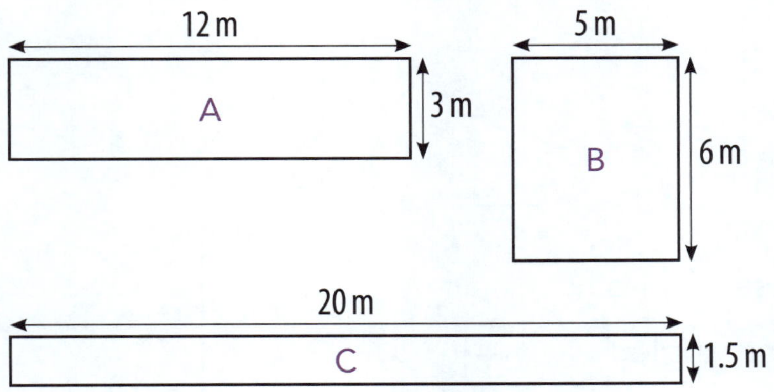

2 (1 mark)

3 Jamie draws a rectangle with an area of **36 cm²**.

What are the lengths of its sides?

Find two possible answers.

______ cm and ______ cm

______ cm and ______ cm

3 (1 mark)

4 How much **larger** is the area of the rectangle than the area of the square? ______ m^2

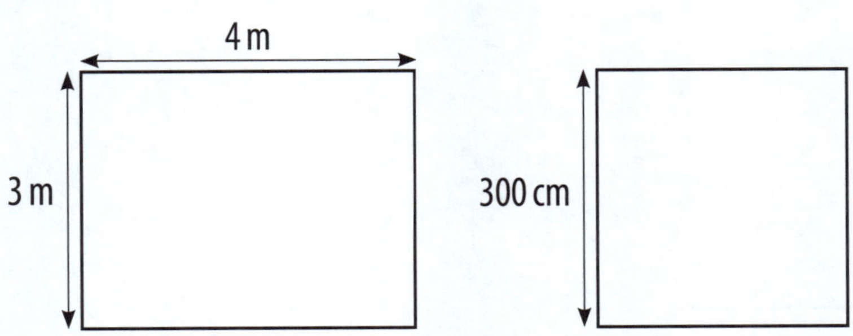

4 (2 marks)

Top tip
- Think about how multiplication is related to area.

/5 Total for this page

Drawing lines and angles

To achieve the expected standard, you need to:

★ complete simple 2-D shapes using given lengths and **acute angles** that are multiples of 5 **degrees**.

1 Draw a rectangle with lengths 6.5 cm and 3.5 cm.

1 (1 mark)

2 A right-angled triangle has a base of 11 cm.

a) Measure the height (x) of the triangle. [ ] cm

b) Measure and label the sizes of each angle.

Angle a = [ ]°

Angle b = [ ]°

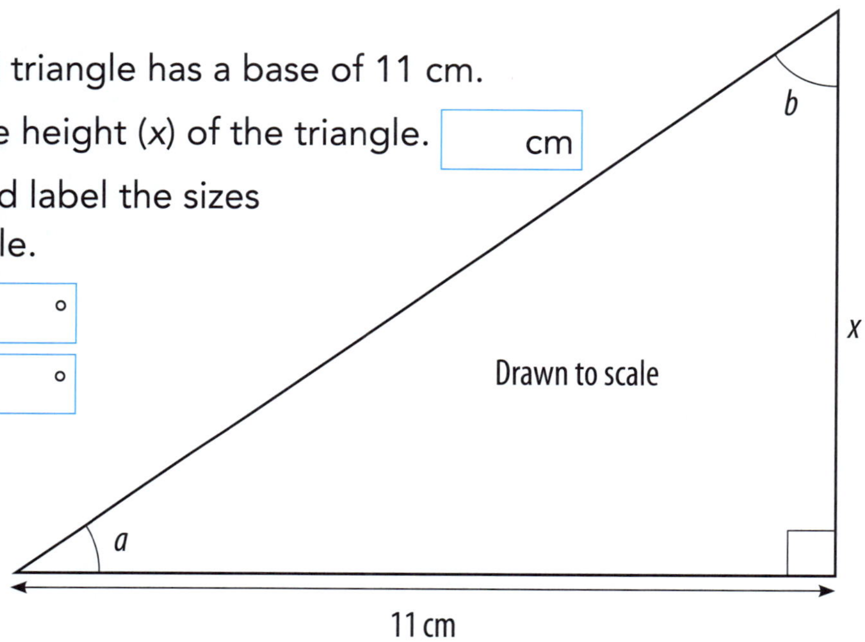

2a (1 mark)

2b (1 mark)

3 a) Use the lines to complete the **trapezium** with angle a = 65°.

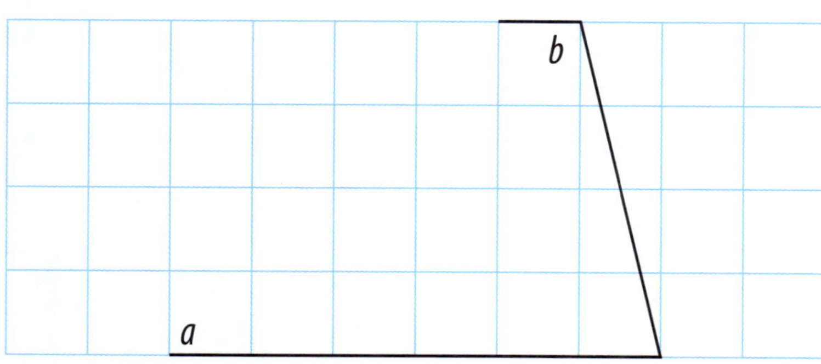

3a (1 mark)

b) Write the size of angle b to the **nearest 5 degrees**. [ ]°

3b (1 mark)

Top tip

- Think how you can be sure that an angle is a right angle.
- Remember that lines or angles on shapes **drawn to scale** can be **measured**.

/5 Total for this page

2-D shapes

To achieve the expected standard, you need to:
- ★ recognise properties of 2-D shapes in terms of number and length of sides, **parallel** sides, type and size of angles
- ★ identify **irregular** and **regular** polygons
- ★ name 2-D shapes according to their properties.

1 Complete the table.

Name of 2-D shape	Number of straight sides	Number of right angles
rectangle		
	3	1
regular hexagon		
equilateral triangle		

(1 mark) 1

2 Tick (✓) all the **hexagons**.

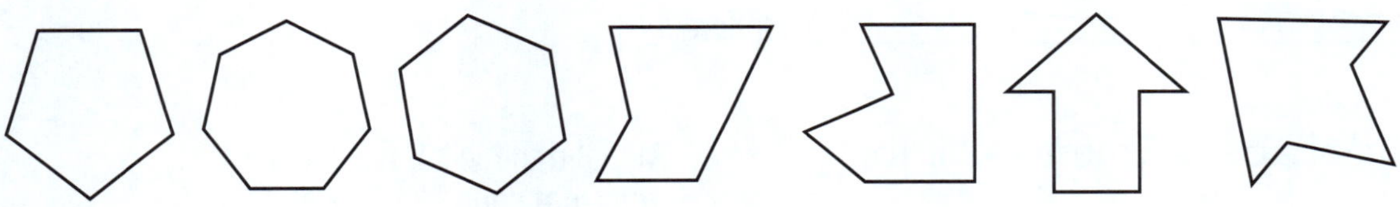

(1 mark) 2

3 Anya wants to draw a shape with **at least** two pairs of **parallel** sides.

Circle **all** the shapes she can draw.

trapezium   rectangle   rhombus   regular pentagon

(1 mark) 3

4 Decide which shape is being described by the table.

Complete the table.

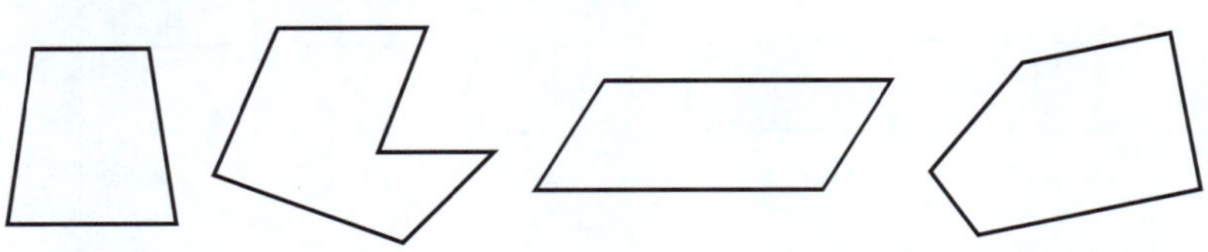

Name of shape	Number of pairs of parallel lines	Number of pairs of perpendicular lines	Total number of sides
	1	2	

(1 mark) 4

Top tip
- Use the corner of a ruler to help identify any acute angles.

/4 Total for this page

3-D shapes

To achieve the expected standard, you need to:

- ★ recognise properties of 3-D shapes – **faces**, **edges** and **vertices**
- ★ recognise and describe simple 3-D shapes
- ★ use **nets** and other 2-D representations of simple 3-D shapes.

1 Fill in the missing information about this 3-D shape.

Number of faces ☐

Number of edges ☐

Number of vertices ☐

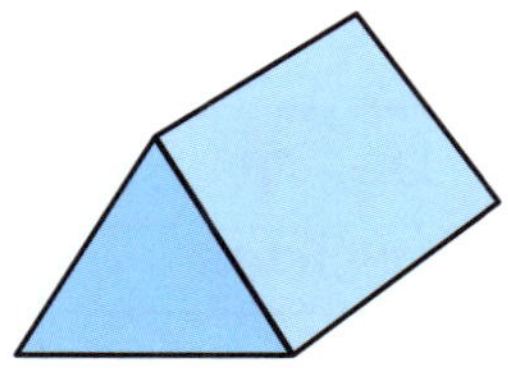

☐ 1 (1 mark)

2 Write the letters of the shapes in the correct place in the Venn diagram.

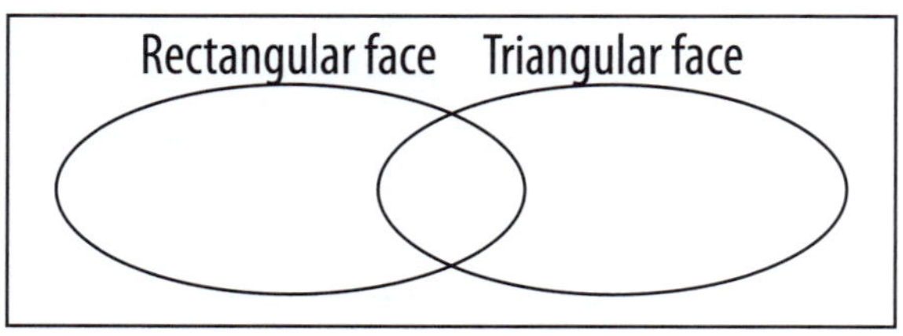

A cone
B triangular prism
C tetrahedron
D cuboid
E sphere

☐ 2 (1 mark)

3 What solid shape is Emma describing?

The solid shape is a ______________________

My solid shape has five faces and eight edges. Four faces are the same shape but one face is square.

☐ 3 (1 mark)

4 Tick (✓) the nets that fold to make a **cube**.

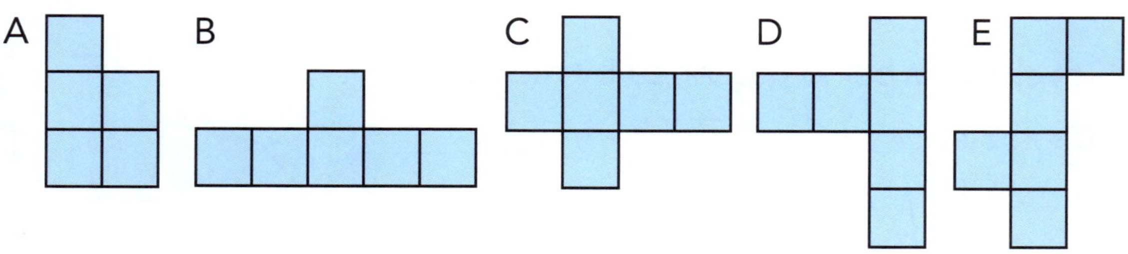

☐ 4 (1 mark)

5 Tick (✓) the nets that fold to make a **triangular prism**.

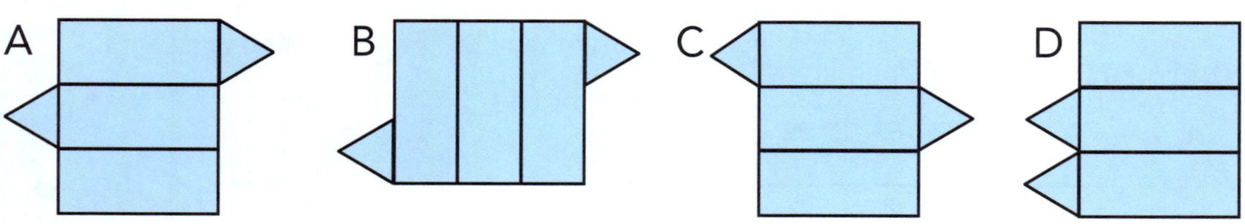

☐ 5 (1 mark)

Top tip

- A picture of a 3-D shape may not show all the faces, edges and vertices.

/5 Total for this page

Angles and degrees

To achieve the expected standard, you need to:

★ recognise angles where they meet at a point or on a straight line
★ find missing angles at a point and along a straight line.

1 Find the size of angle *m*. [ ]°

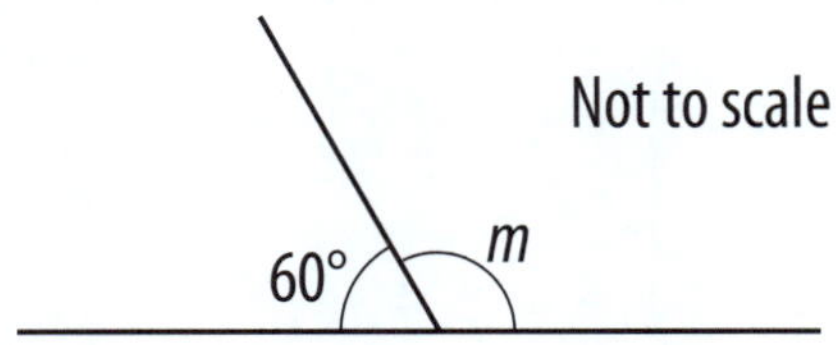

(1 mark) 1

2 Find the size of angle *a*. [ ]°

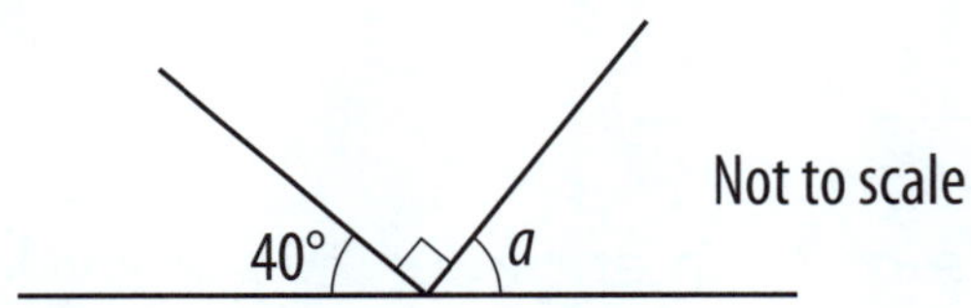

(1 mark) 2

3 Calculate angles *a* and *b*.

Angle *a* = [ ]°

Angle *b* = [ ]°

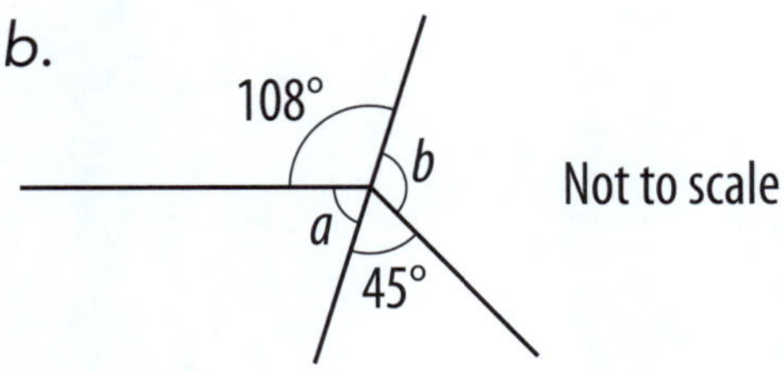

(1 mark) 3

4 Calculate the size of angle *p*. [ ]°

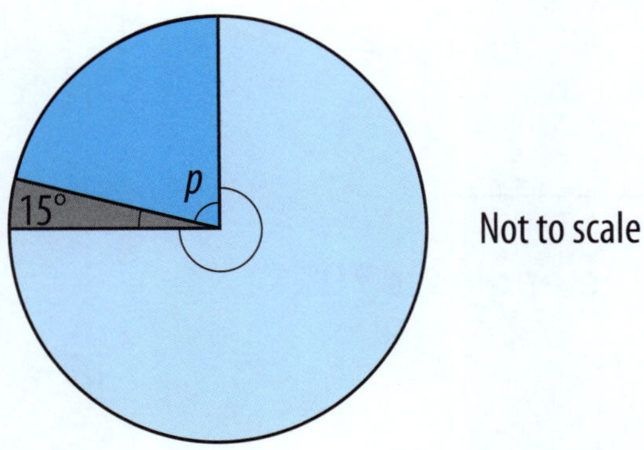

(1 mark) 4

5 Calculate the size of angle *q*. [ ]°

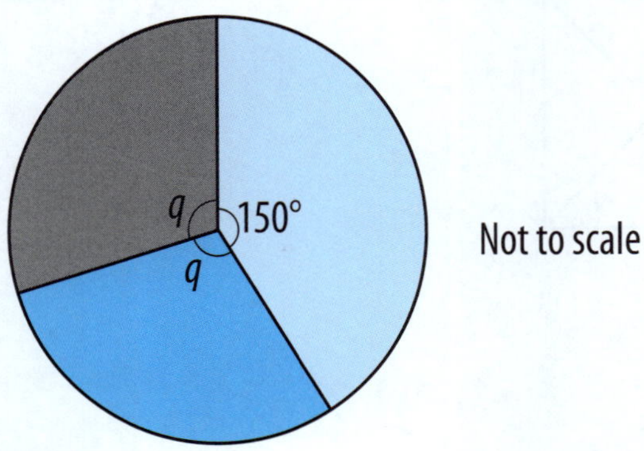

(2 marks) 5

Top tip

- Look out for the symbol that represents a right angle in missing angle questions.

/6

Total for this page

Angles in triangles

To achieve the expected standard, you need to:
★ find unknown angles in different types of triangles.

1 Calculate the size of the angle *m*. [ ]°

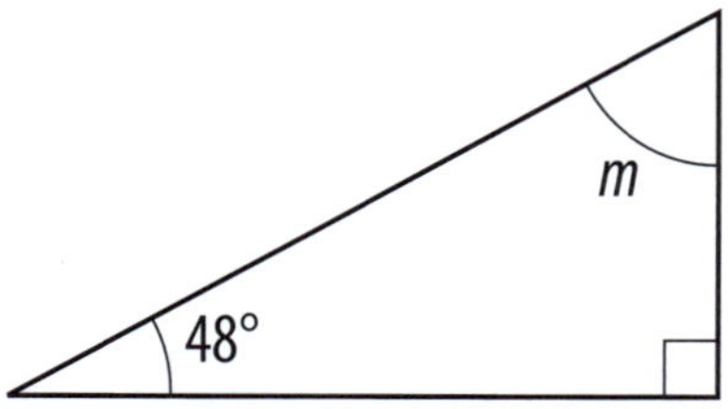

1 (1 mark)

2 Write the size of **each** angle on the triangle.

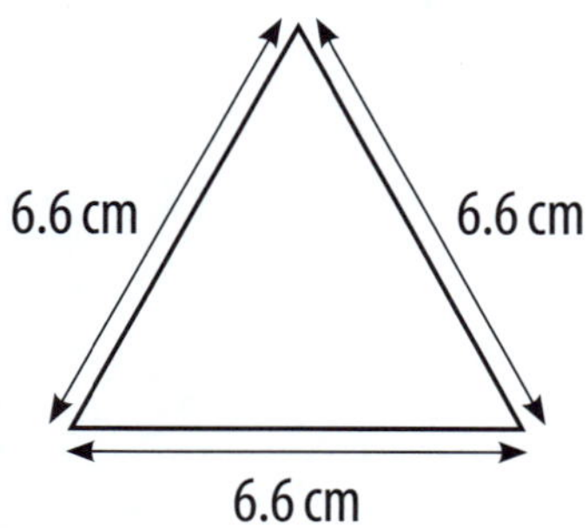

[ ]°

2 (1 mark)

3 Work out the size of angles *a* and *b*.

Angle *a* = [ ]°

Angle *b* = [ ]°

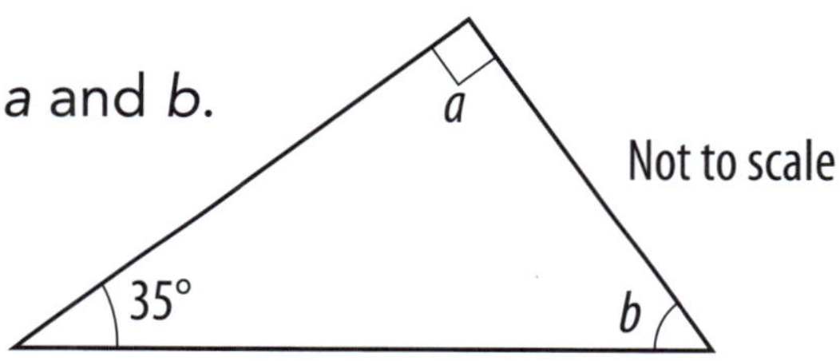

3 (1 mark)

4 Alfie draws an isosceles triangle. Only **one** of the angles is **70°**.

What are the sizes of the other angles?

[ ]° and [ ]°

4 (1 mark)

5 Calculate the sizes of angles *a* and *b*.

a = [ ]°

b = [ ]°

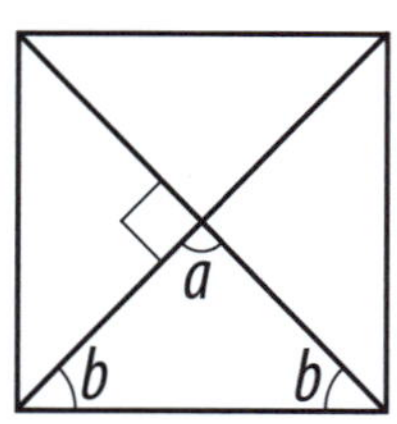

5 (1 mark)

Top tip

- Remember that an isosceles triangle has two angles the same size, as well as two sides of the same length.

/5 Total for this page

Coordinates

To achieve the expected standard, you need to:

- ★ describe the positions on a 2-D coordinate grid using axes with equal scales in the first quadrant and become more confident in plotting points in all four quadrants
- ★ use **coordinates** to complete a given rectangle.

1 Write the coordinates of each vertex of the triangle.

A (,)

B (,)

C (,)

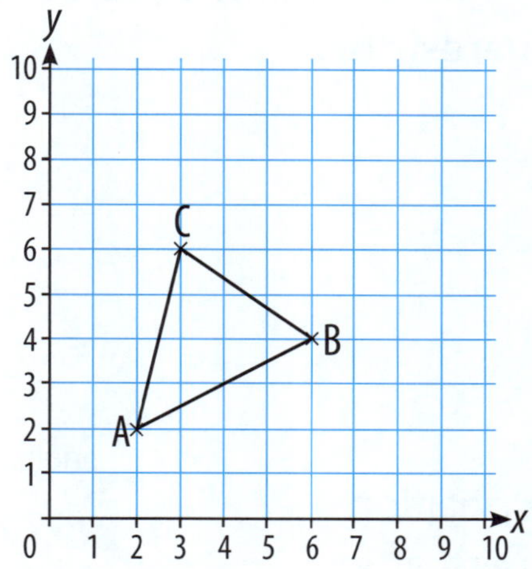

(1 mark) 1

2 Write the coordinate of vertex **D** that completes the rectangle.

D (,)

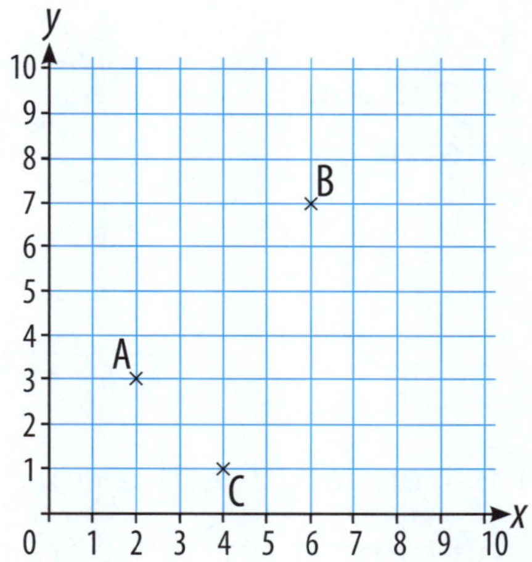

(1 mark) 2

3 Plot these coordinates on the grid.

A (1,0) **B** (–2,2) **C** (–2,7) **D** (2,7) **E** (4,4)

Join the coordinates in order to make a closed shape.

Name the shape: ____________________

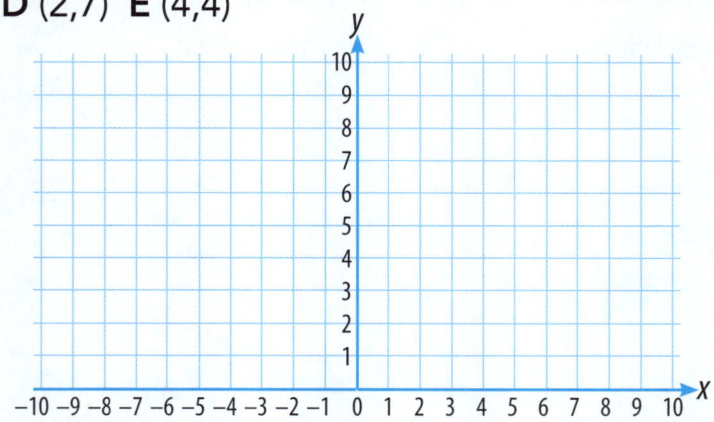

(2 marks) 3

4 Jayden plots (5,4) and (8,4) as two vertices of a square.

What could the coordinates of the other corners be?

(,) and (,)

(1 mark) 4

Top tip

- Coordinates are always written in the order (x,y).

/5 Total for this page

Translations

To achieve the expected standard, you need to:

★ identify, describe and represent the position of a shape following a **translation**.

1 Write the **new** position of coordinate **M** after the translation **2 squares left** and **3 squares down**.

M (,)

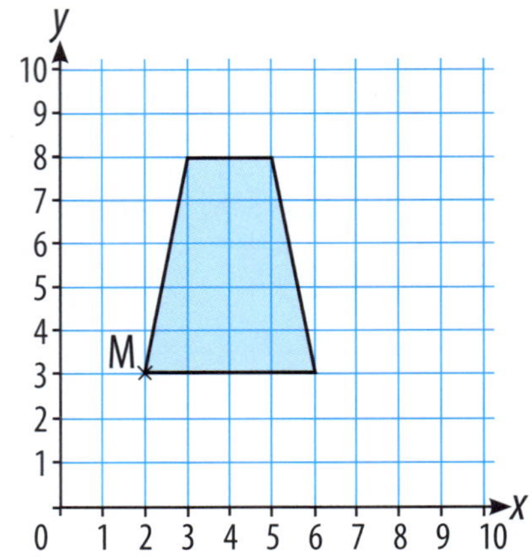

(1 mark)

2 Draw the position of the square after the translation **8 squares right** and **2 squares down**.

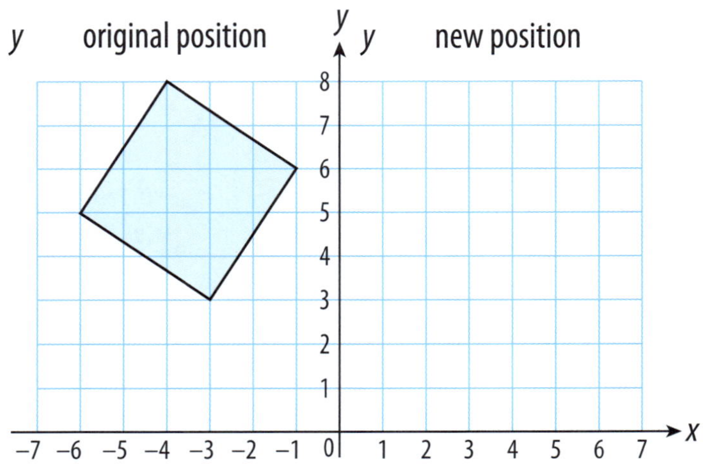

(1 mark)

3 Triangle 1 is translated to the position of Triangle 2.

Describe the translation of the triangle.

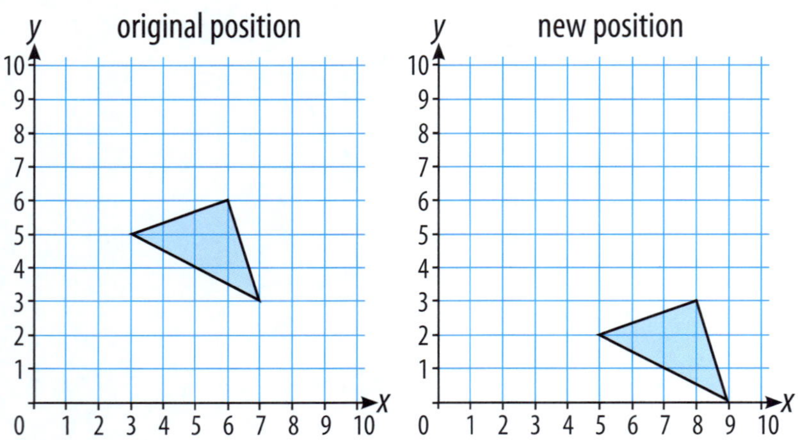

(1 mark)

Top tip

- Imagine a translation as a shape sliding horizontally and vertically. The size and orientation of the shape do not change, just the position.

/3

Total for this page

Reflections

To achieve the expected standard, you need to:
★ identify, describe and represent the position of a shape following a **reflection**.

1 Tick (✓) the shape that has been **correctly** reflected in the mirror line.

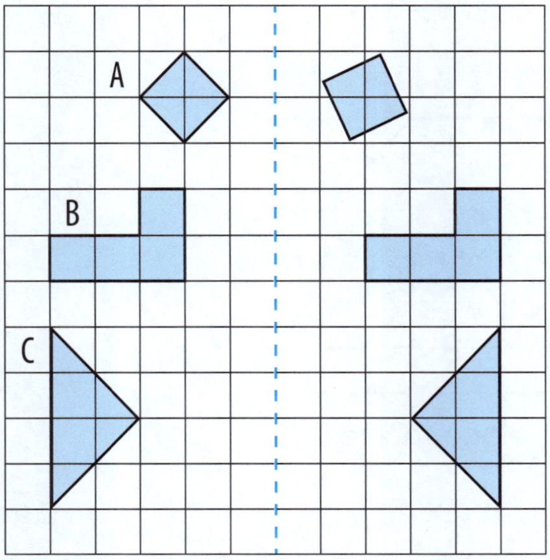

mirror line

(1 mark) 1

2 Reflect the shape in the mirror line.

Draw the reflected shape.

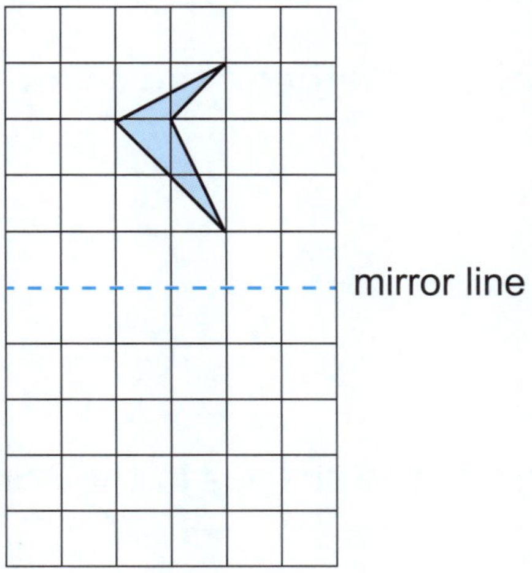

(1 mark) 2

3 Reflect the shape in the mirror line.

Draw the reflected shape.

mirror line

(1 mark) 3

4 Reflect the shape in the **x axis**.

Draw the reflected shape.

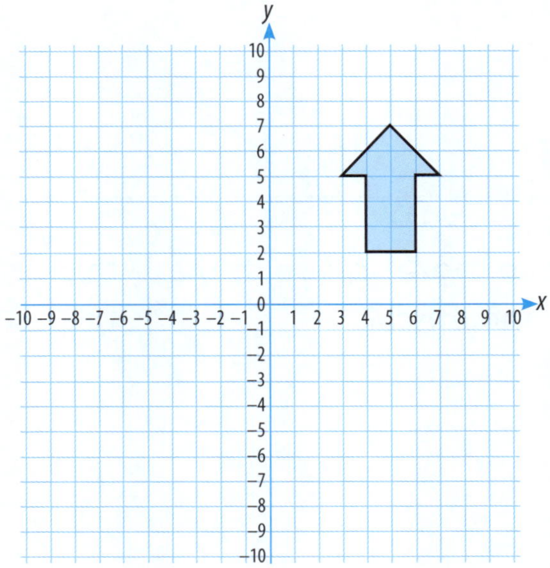

(1 mark) 4

Top tip
- Shapes do not change size when they are reflected.

/4 Total for this page

Reflective symmetry

To achieve the expected standard, you need to:

★ compare and classify 2-D shapes in terms of **reflective symmetry**.

1 Tick (✓) the shapes that have reflective symmetry.

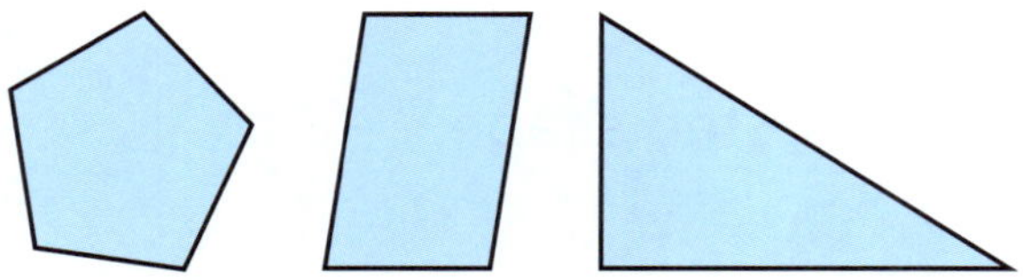

(1 mark)

2 Draw the line of symmetry on the shape.

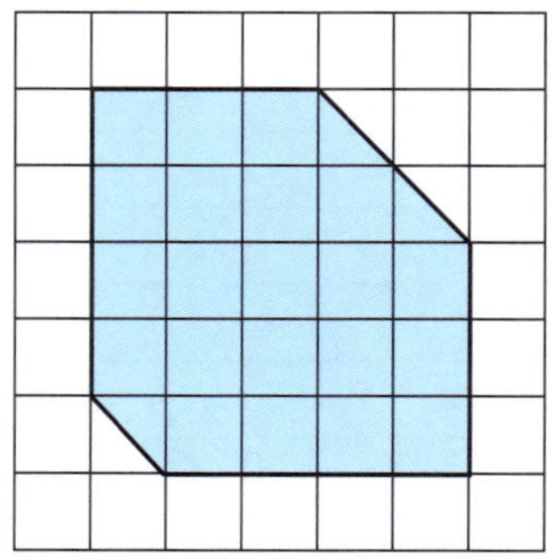

(1 mark)

3 Write the letters of the shapes in the correct place in the Carroll diagram.

	Reflective symmetry	No reflective symmetry
Quadrilateral		
Not quadrilateral		

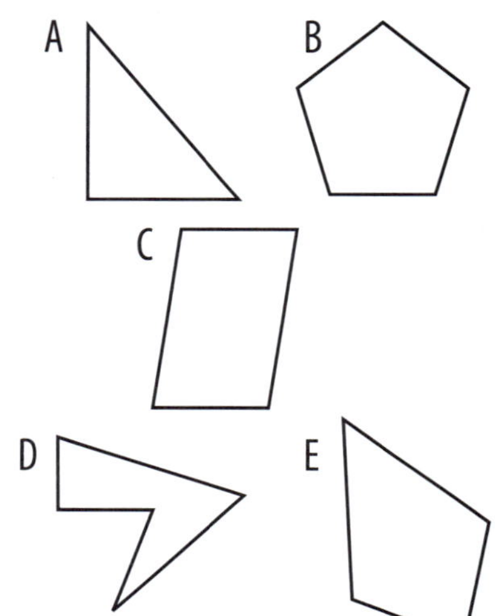

(1 mark)

4 Shade **one more** square to make this shape symmetrical.

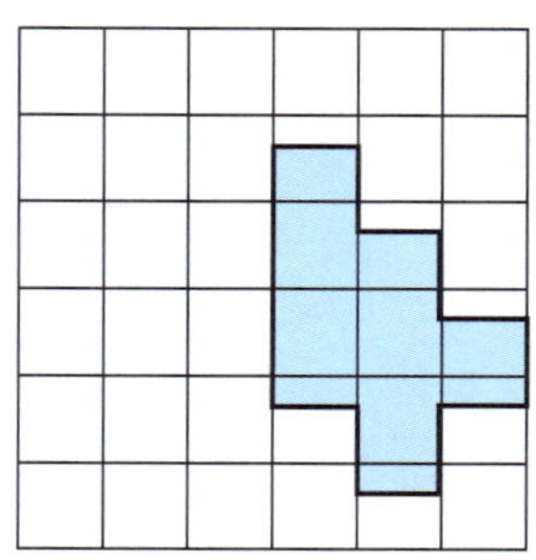

(1 mark)

Top tips

- Use a real mirror to help you. Draw mirror lines using a ruler.
- It is often useful to turn the paper so that you can look at a shape in different ways.

/4

Total for this page

Tables

To achieve the expected standard, you need to:
★ complete, read and interpret information in **tables**.

1 Complete the table showing the children's favourite sport.

Favourite sport	Tally	Total
badminton	卌 卌 卌 卌 III	
hockey	卌 卌 II	12
football	卌 卌 卌 III	

(1 mark) 1

2 a) How much money have the children saved in total? ______

	Sally	Ash	Peta	Pav
Money saved	£7.75	£10.99	£9.20	£11.49

(1 mark) 2a

b) How much **more** has Pav saved than Sally? ______

(1 mark) 2b

3 The table shows the days of the week the players are free to play a tennis match.

	Monday	Tuesday	Wednesday	Thursday	Friday
Kate	✓		✓		✓
Zack		✓			✓
Jayden		✓		✓	
Dina	✓			✓	✓

a) On which day of the week can **no** matches be played?

______

(1 mark) 3a

b) Which two players are **not** able to have a match together?

______ and ______

(1 mark) 3b

4 This table shows how much water was collected in a week in three different containers.

	Container A	Container B	Container C
Amount of water container holds	$3\frac{1}{2}$ litres	$4\frac{1}{4}$ litres	$2\frac{3}{4}$ litres
Number of containers filled	4	3	2

a) How much **more** water can container B hold than container C? ______ litres

(1 mark) 4a

b) How many litres have been collected **altogether**? ______ litres

(2 marks) 4b

/8 Total for this page

Pictograms

To achieve the expected standard, you need to:

★ interpret and present data using **pictograms**.

1 Complete the pictogram using the information shown in the table.

	Total points
Red	12
Blue	16
Green	20

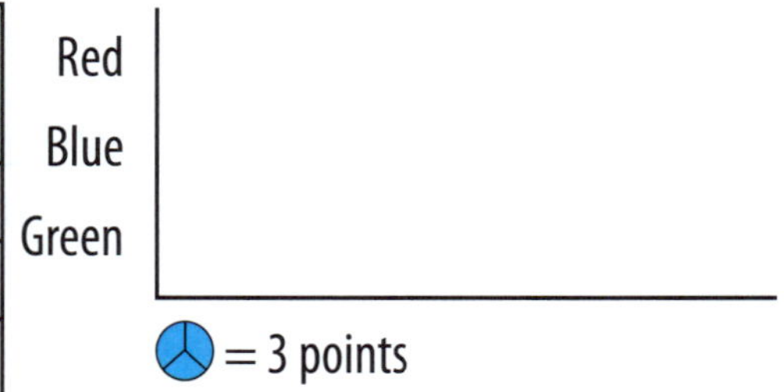

1 (1 mark)

2 a) How many **more** children like canoeing than treasure hunts?

b) How many children like outdoor activities in **total**?

c) Nine children do **not** like any outdoor activities.

Show this on the pictogram using the same symbols.

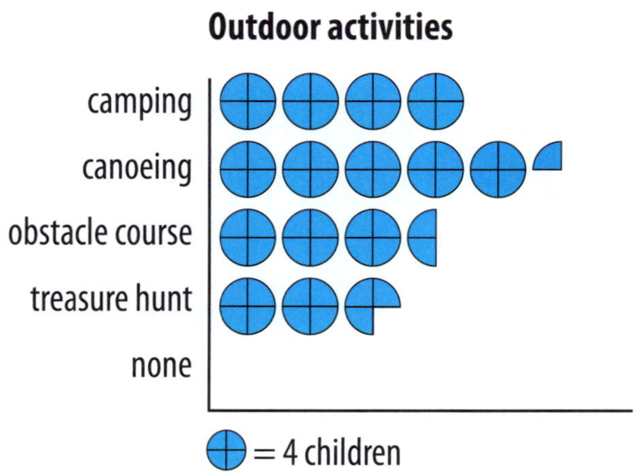

2a (1 mark)

2b (1 mark)

2c (1 mark)

3 a) Which magazines sold **fewer** than 25 copies?

b) Which two magazines sold a **total** of 52 copies?

and

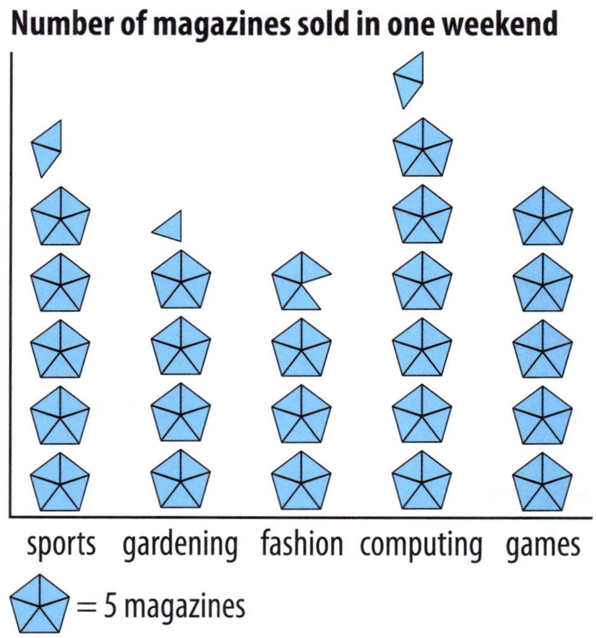

3a (1 mark)

3b (1 mark)

Top tip

- Check the value of each part of the pictogram before tackling the question.

/6 Total for this page

Bar charts

To achieve the expected standard, you need to:
★ complete, read and interpret information presented in **bar charts**.

1 This bar chart shows the types of library books borrowed by Class 6 last term.

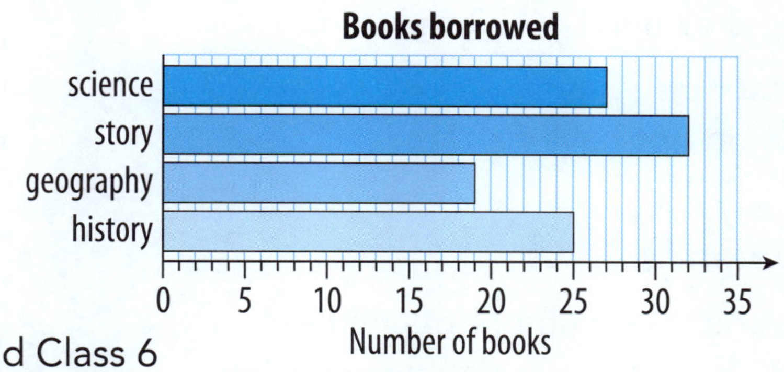

a) How many **more** story books than geography books were borrowed? ☐

1a (1 mark)

b) How many books did Class 6 borrow in **total** from the library? ☐

1b (1 mark)

2 This bar chart shows the mass of different objects.

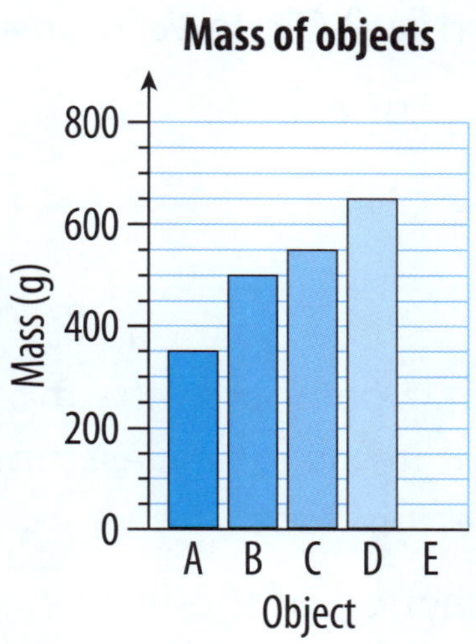

a) Complete the table using the bar chart to help you.

	Object A	Object B	Object C	Object D
Mass (g)				

2a (1 mark)

b) Object E has a mass of 450 g. Draw the missing bar on the chart.

2b (1 mark)

3 This bar chart shows the money raised by each class in a cake sale.

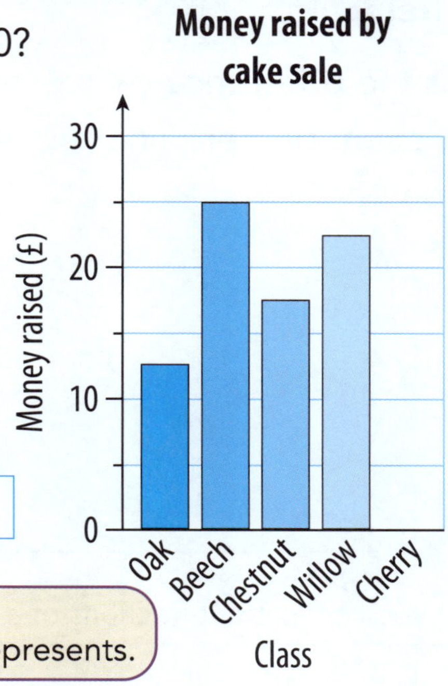

a) How many classes raised **more** than £20? ☐

3a (1 mark)

b) Cherry Class raised £1.25 **less than** Oak Class. Draw the missing bar on the chart.

3b (1 mark)

c) The school cake sale target was **£100 altogether**.

How much **more** money did they need to raise to reach the target? ☐

3c (2 marks)

Top tip

- Look at the scale to see what each interval represents.

/8 Total for this page

Pie charts

To achieve the expected standard, you need to:
★ interpret simple **pie charts**.

1 The pie chart compares the different types of small pets owned by 120 children.

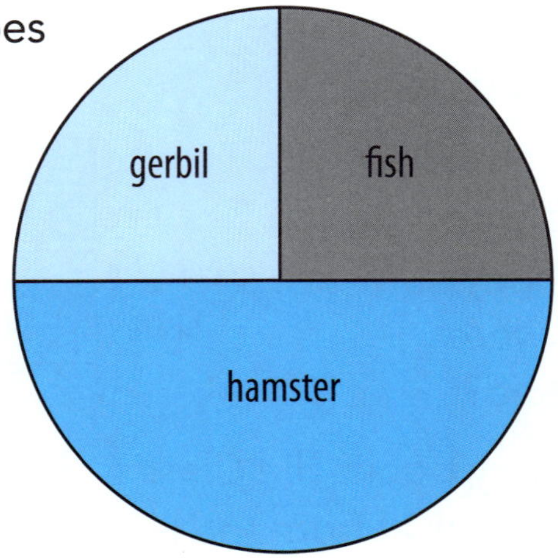

a) What **fraction** of the children do **not** have **fish**? [ ]

b) How many children do **not** have fish? [ ]

c) Ed says the total number of children with fish and gerbils is **equal** to the number of children with hamsters.
Is he correct? Circle your answer. YES / NO

Explain your answer.

____

____

1a (1 mark)
1b (1 mark)
1c (1 mark)

2 This pie chart compares the money spent on different items for a classroom. £240 is spent in total.

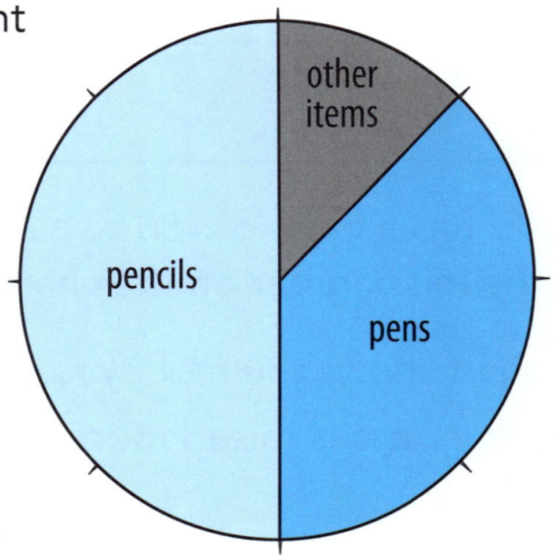

a) What **fraction** of the total amount is spent on pens? [ ]

b) How much money is spent on **other items**? [ ]

c) What is the **difference** between the money spent on pens and the money spent on pencils? [ ]

2a (1 mark)
2b (1 mark)
2c (1 mark)

Top tip

- Use what you know about fractions of a circle to help with pie chart questions.

/6
Total for this page

Line graphs

To achieve the expected standard, you need to:
★ interpret **line graphs** to solve problems.

1 The graph shows the growth of a bean plant.

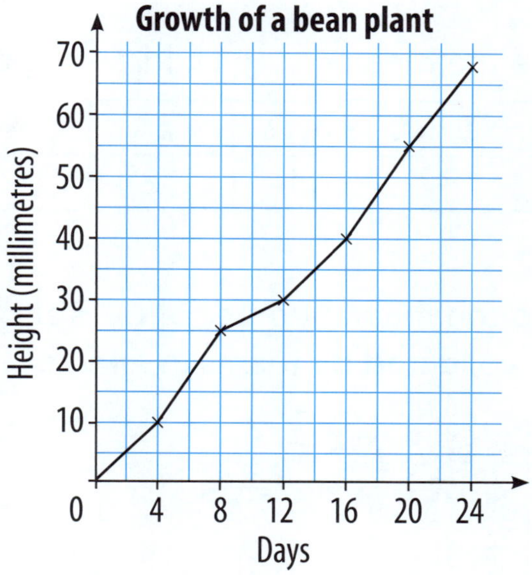

a) Use the graph to help you complete the table.

Day	4	8	14	20
Height (mm)				

b) Estimate the height of the bean plant on **day 24**. ______ mm

1a (1 mark)

1b (1 mark)

2 This graph shows the temperature outside at different times of the day.

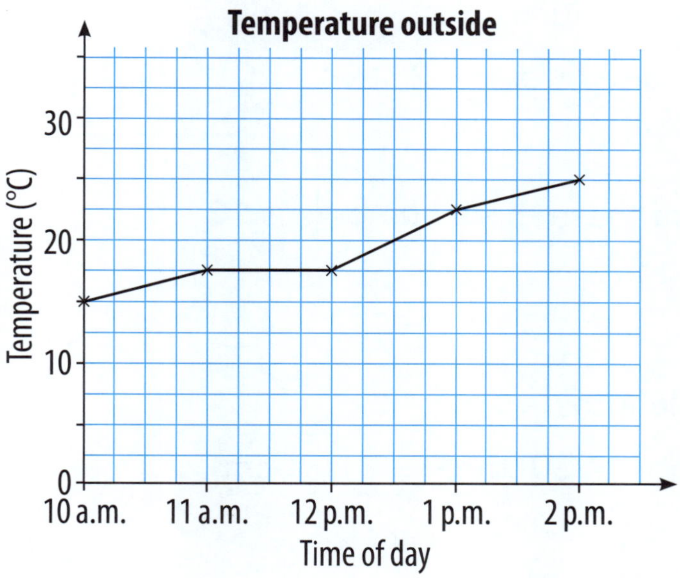

a) How much warmer is the temperature at **1 p.m.** than at **11 a.m**?
______ °C

b) What was the temperature at **12:30 p.m**? ______ °C

c) For how long was the temperature **below 20°C**?

______

2a (1 mark)

2b (1 mark)

2c (1 mark)

Top tip
- Look carefully at the intervals and scales on each graph.

/5
Total for this page

Mean

To achieve the expected standard, you need to:

★ calculate and interpret the **mean** as an **average** for simple sets of discrete data.

1 Calculate the mean of the children's test scores.

Oscar	Harriet	Lewis	Katy	Theo
9	7	11	12	6

1 (1 mark)

2 Calculate the mean number of buckets of water collected by teams in a race.

2 (1 mark)

3 Find the mean of these capacities.

$1\frac{1}{2}$ litres 3 litres $3\frac{1}{2}$ litres 2 litres

litres

3 (1 mark)

4 15 11 ? The **mean** of these three numbers is **12**

What is the missing number?

4 (1 mark)

5 Find the mean of these lengths.

90 cm 60 cm 0.5 m 30 cm

cm

5 (1 mark)

6 Calculate the average number of **minutes** per weekday that children spend playing sports.

minutes

Monday	Tuesday	Wednesday	Thursday	Friday
30 minutes	1.5 hours	45 minutes	2 hours	35 minutes

6 (2 marks)

Top tip

- Always check the unit of measurement, and convert any that are in a different unit.

/7

Total for this page

Answers

Number and place value

Place value of whole numbers (page 6)

1 9 hundred (900), 9 thousand (9,000), 9 tens (90), 9 hundred thousand (900,000)

2 Award **2 marks** for all four correct
Award **1 mark** for any three correct

7 thousand — 597,302
900 thousand — 979,200
30 thousand — 1,233,420
3 million — 3,095,799

3 Any two numbers that use all the digit cards to make two numbers between 5,750 and 6,250 (e.g. 5,780 and 6,134)

4 3,463,400 and 3,483,400

5 1,150,000 (1,005,000)
(1,050,000) 1,105,000

Comparing and ordering whole numbers (page 7)

1 a) 34,601 > 34,599
b) 709,898 < 709,988
Award **1 mark** for both answers correct.

2 940,500 g 904,500 g 95,400 g 94,500 g 9,500 g

3 a) 4,801 b) 4,090

4 All years must be ordered correctly.

	Year
Greatest number of visitors	2015
	2013
	2012
	2016
	2017
Fewest number of visitors	**2014**

Rounding (page 8)

1 4 for ☐78 and 0; 1; 2; 3; or 4 for 5☐0

2 a) 130 b) 100 c) 540
Award **1 mark** for all answers correct.

3 952; 1,045; 1,009

4 Award **2 marks** for all correct.
Award **1 mark** for at least six correct.

	To the nearest 10	To the nearest 1,000	To the nearest 10,000
45,419	45,420	45,000	50,000
234,549	234,550	235,000	230,000
3,126,095	3,126,100	3,126,000	3,130,000

5 a) 5,745 b) 5,749

Place value of decimal numbers (page 9)

1 10

2 8.3; 7.3; 6.8

3

2.5	4 or 4.0	5.5	7 or 7.0	8.5	10 or 10.0

4 8.04

5 Ben is incorrect because 6 equals 6 ones and 0.06 equals 6 hundredths, which is less than 6 ones.

6

2.7	7.02	7.2	20.7	70.02
smallest				**largest**

Negative numbers (page 10)

1 1; –5

2 5 levels

3 –7

4 a) Two different pairs, e.g. 10 and –1, 9 and –2, 8 and –3, 7 and –4, 6 and –5, 5 and –6
b) Award **1 mark** for a response of NO, and an explanation to show that either:
- 11 and 0 have a difference of 11 but 0 is not a negative number
- one of the numbers has to be negative and 0 is not negative (implied that 0 and 11 have the required difference).

Do not award any marks for only identifying 11 and 0

Number – Addition, subtraction, multiplication and division

Addition (page 11)

1 800

2 121,483

3 a) 15,476
b) 22,074

4

	3	9	4	8
+	2	**1**	4	5
	6	0	**9**	**3**

5 Award **2 marks** for correct answer £550,696
Award **1 mark** for evidence of appropriate working but with an incorrect answer.
Do not award any marks if a final answer is missing.

Subtraction (page 12)

1 10,924

2 160,000

3 29,138

4 Award **2 marks** for the correct answer of £5,122
Award **1 mark** for evidence of the correct method with no more than one mathematical error, e.g.
£2,759 + £5,764 + £1,905 = £10,428
£15,550 – £10,428 = wrong answer

5

	4	0	**8**	8	6
–	2	**4**	6	3	**7**
	1	6	2	**4**	9

Multiplying and dividing by 10 and 100 (page 13)

1 a) 23,500 b) 100 c) 0.99

Award **1 mark** for all answers correct.

2 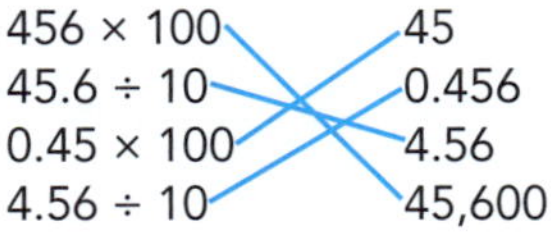

3 Award **2 marks** for all three correct.

Award **1 mark** for any two correct.

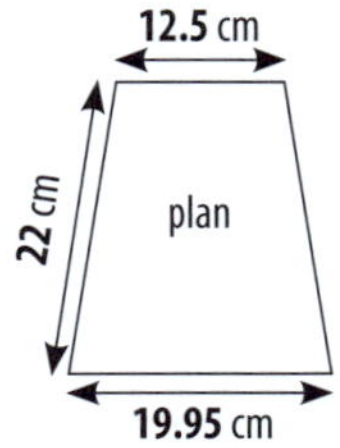

4 Parcel A is 25 kg and Parcel B is 2.5 kg.

Multiples and factors (page 14)

1 77; 220; 1,100

2 1 or 3

3 **2 marks** for all numbers positioned correctly; **1 mark** for 4 out of 5 numbers positioned correctly.

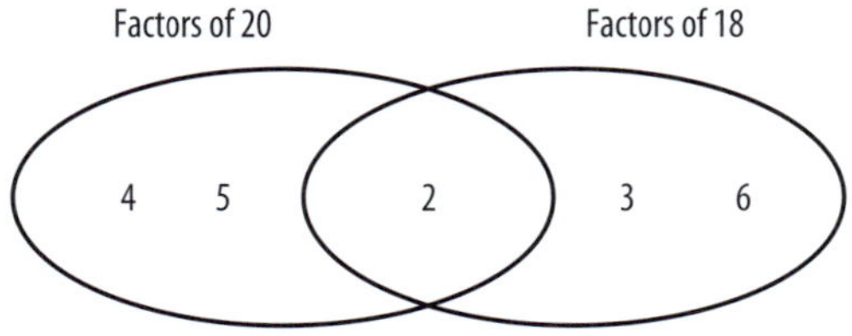

4 1; 2; 4; 8; 16; 32

5 2; 6; 10; 30

Multiplying by larger numbers (page 15)

1 25,263

2 99,935

Award **2 marks** for the correct answer.

Award **1 mark** for evidence of correct method with no more than one mathematical error.

3 8,925

4 a)

	2	5	**6**
×		2	6
1	**5**	3	6
5	1	2	**0**
6	6	5	6

b)

	2	**0**	8
×		**3**	5
1	0	4	0
6	2	4	0
7	2	**8**	0

5 Award **2 marks** for the correct answer of 14,976.

Award **1 mark** for evidence of the correct method with no more than one mathematical error, e.g.

52 × 24 = 1,248

1,248 × 12 = wrong answer

Square numbers (page 16)

1 25 and 16

2 64

3 e.g. 1 + 4 = 5; 4 + 9 = 13; 1 + 16 = 17; 4 + 25 = 29

4 Amy

5 Award **1 mark** for a response of YES, and an explanation to show that either:

- 81 and 100 are the only two numbers that are square between 80 and 110
- 9^2 is 81 and 10^2 is 100, so these are the only two square numbers between 80 and 110.

Short division (page 17)

1 62

2 638

3 884

4 44

5 477.5 or $477\frac{1}{2}$

Long division (page 18)

1 Award **2 marks** for correct answer 57.

Award **1 mark** if answer is incorrect, but there is evidence of using the formal method of long division. Working must be carried through to reach an answer for the award of 1 mark. In all cases accept follow-through of one error in working, e.g.

wrong answer

```
15 | 8 5 5
     7 5 ↓
     1 0 5
     1 0 5
         0
```

Do not award any marks if a final answer is missing.

2 Award **2 marks** for correct answer 47

Award **1 mark** if working has been carried through to reach an answer and accept follow-through of one error (see Q1 above).

Do not award any marks if a final answer is missing.

3 33

4 Award **2 marks** for correct answer £65

Award **1 mark** if the answer is incorrect, but there is evidence of an appropriate method.

5 Award **2 marks** for correct answer 21 (there is not enough ribbon to make 22).

Award **1 mark** if the answer is incorrect, but there is evidence of an appropriate method.

Prime numbers (page 19)

1 2; 7; 11; and 17

2 3

3 2 and 17

4 2; 3; or 5

5 6 groups of 7 children indicated only.

6 2 + 11 + 13 **and** 2 + 7 + 17 **and** 2 + 5 + 19

Award **2 marks** for **three sets** of correct responses.

Award **1 mark** for **two sets** of correct responses.

Do not award any marks for only one set of correct responses.

Number – Fractions, decimals and percentages

Fractions of amounts (page 20)

1 a) 5 b) 15

Award **1 mark** for both answers correct.

2 £30 (Megan has £90, Ali has £60)

3 3

4 a) £600 b) £3,600

5 $\frac{3}{5}$

6 Award **2 marks** for the correct answer of 108 pages.
Award **1 mark** for evidence of the correct method with no more than one mathematical error, e.g.
72 pages ÷ 2 = 36 pages
36 pages + 72 pages = incorrect answer

Mixed numbers (page 21)

1 $\frac{20}{3}$

2 $1\frac{1}{4}$, $1\frac{3}{4}$, $2\frac{1}{4}$, $2\frac{3}{4}$, $3\frac{1}{4}$

3 $1\frac{1}{2}$ or $1\frac{4}{8}$

4 Award **2 marks** for all correct.
Award **1 mark** for any three correct.

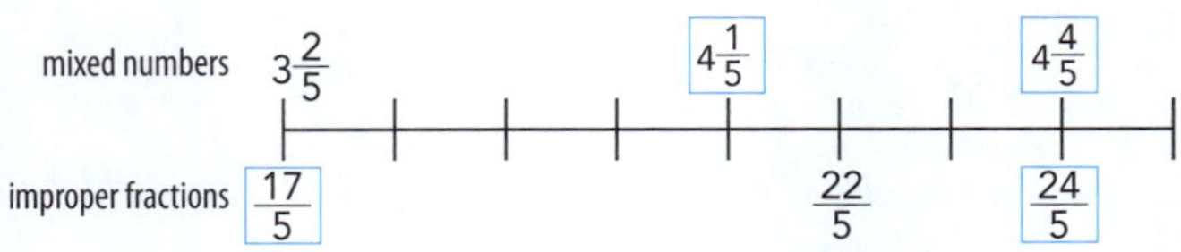

5 $\frac{30}{5}$ $\frac{18}{3}$ $\frac{54}{9}$

6 $2\frac{3}{4}$ or $\frac{11}{4}$

Equivalent fractions (page 22)

1
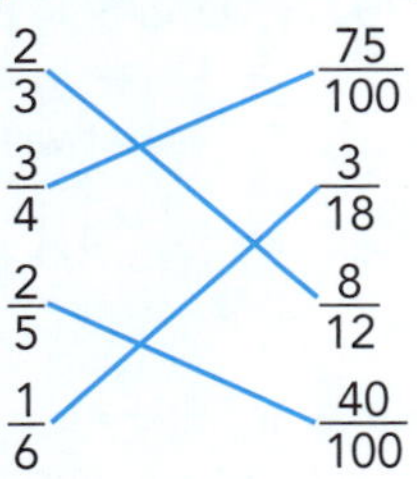

2 $\frac{15}{25}$

3 $\frac{1}{4}$ and $\frac{3}{4}$

4 $\frac{4}{100}$m and $\frac{1}{4}$m

5 $\frac{10}{20}$ $\frac{8}{20}$ $\frac{15}{20}$

6 $\frac{5}{8}$ (five triangles) shaded in any way

Adding and subtracting fractions (page 23)

1 $1\frac{2}{5}$

2 $\frac{4}{5}$

3 $\frac{9}{6}$ **or** $1\frac{3}{6}$ **or** $1\frac{1}{2}$

4 $\frac{3}{4}$ or any equivalent

5 Award **2 marks** for all correct.
Award **1 mark** for at least two correct.

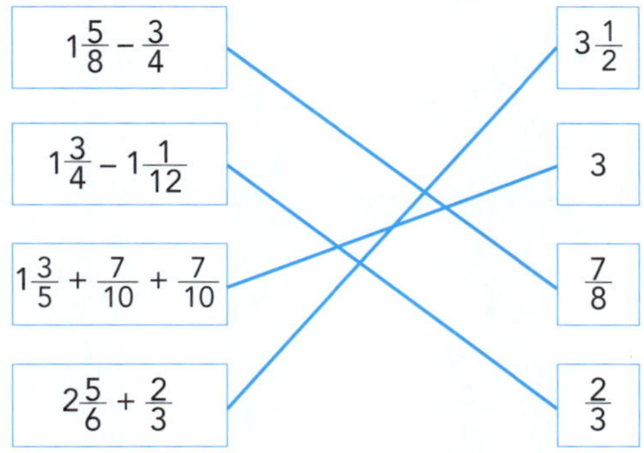

6 $1\frac{1}{8}$

Fractions and their decimal equivalents (page 24)

1 Award **2 marks** for all correct.
Award **1 mark** for three correct.

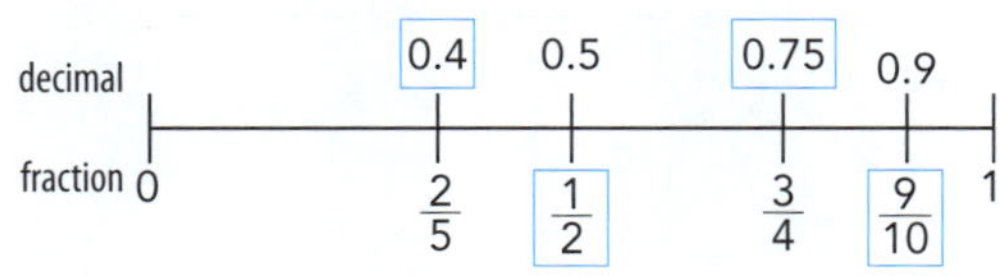

2 $\frac{36}{100}$

3 $\frac{25}{100}$, $\frac{5}{20}$ and $\frac{1}{4}$

4
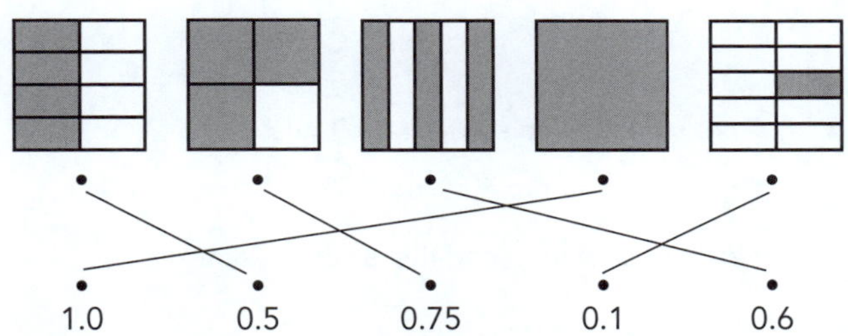

Adding and subtracting decimals (page 25)

1 a) 44.15
b) 3.9
Award **1 mark** for both answers correct.

2 a) £35.24
b) £4.76

3 1.7 and 4.3; 2.46 and 3.54

4 2.11

5 Containers B and C (also award the mark for calculations that clearly show that B and C have been identified but have not been written in the answer boxes).

Multiplying decimals (page 26)

1 0.7

2 6.4

3 2

4 4.7

5

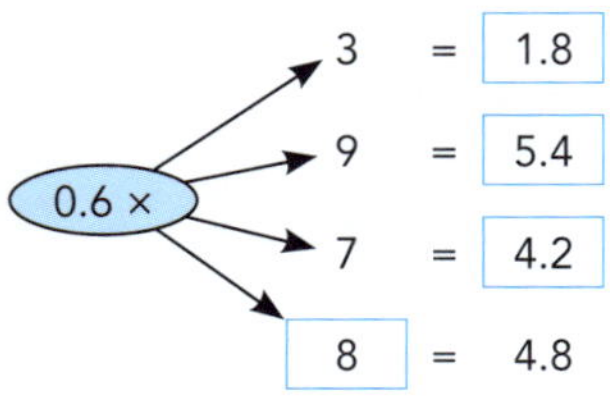

Award **2 marks** for all four correct or **1 mark** for three correct.

6 0.8 (accept $\frac{4}{5}$)

7 12.5 m^2

Percentages as fractions and decimals (page 27)

1 Award **2 marks** for all correct.

Award **1 mark** for at least four correct.

Fraction	$\frac{3}{10}$	$\frac{65}{100}$ or $\frac{13}{20}$	$\frac{5}{100}$ or $\frac{1}{20}$	$\frac{4}{5}$
Decimal	0.3	0.65	0.05	0.8
Percentage	30%	65%	5%	80%

2 1.4; 14%; 2.5

3 $\frac{7}{10}$, 70, 0.7

4 Award **2 marks** for all correct.

Award **1 mark** for at least three correct.

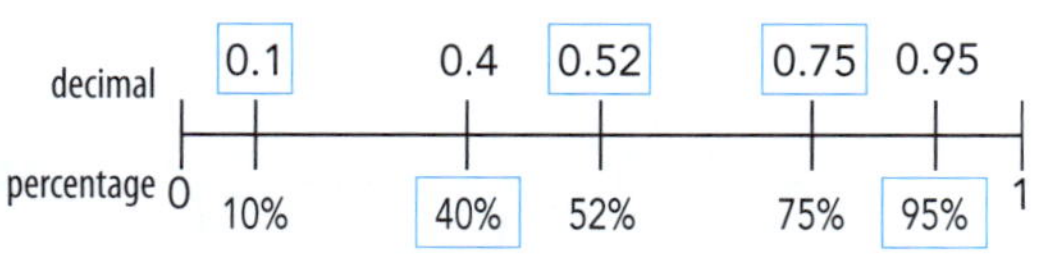

5 a) An explanation to show that either:

- $\frac{2}{5}$ is equivalent to 40%, not 25%
- 25% is equal to $\frac{1}{4}$ and this is not the same as (or is smaller than) $\frac{2}{5}$

b) 0.4

Finding percentages (page 28)

1 a) 24 b) 50

2 a) 200 b) 400

3 a) £40 b) 20 kg

4 =

5 a) Award **2 marks** for 80 shown clearly on the bar chart.

Award **1 mark** for a bar clearly identified but drawn incorrectly or missing.

Do not award any marks for bar shown as 60 or less or 100 or more with no evidence of 80 identified in some way.

b) 168

Ratio and proportion

Ratio and proportion (page 29)

1 27

2 400

3 Award **2 marks** for correct answer 72

Award **1 mark** for the scaling by 12 completed correctly, but with a mis-measurement and an answer (e.g. mis-measured as 5 cm and answer given as 60 km).

4 Accept $\frac{17}{20}$, 17 out of 20, 85% or 0.85

5 Red Team 100, Blue Team 60

Award **2 marks** for both answers correct.

Award **1 mark** for workings that clearly show 160 ÷ 8 = 20 as part of the calculation **and** one team's score is correct.

Algebra

Algebra (page 30)

1 93 in top row, 5 in bottom row

2 36; 35; 13

3 £16.50

4 180 – (B + C)

5 a) 48

b) $q = 20$

Sequences (page 31)

1 –8; –1; 6; 13; 20

2 The rule is subtract 20. Missing terms are 172, 132, 112

3 148 **138** 128 **118** 108; the rule is subtract 10

4 –27

5 Award **1 mark** for NO, and an explanation that shows that:

- since the sequence begins at 35 and adds 25, a multiple of 25 (725) will not be in the sequence.
- 710 and 735 would be in the sequence and 725 is between these numbers.
- the sequence always adds 35 to a multiple of 25, so 700 + 35 = 735 and / or 725 + 35 = 760

Do **not** award any marks for recognition that all numbers in the sequence will end in 00, 25, 50 or 75 as this could imply that multiples of 25 are being counted.

Solving equations (page 32)

1 e.g. 3 + 15 = 6 × 3

2 25 minutes and 35 minutes

3 a) 20 b) 15

4 A = 750 g, B = 250 g

Measurement

Length (page 33)

1 325

2 18

3

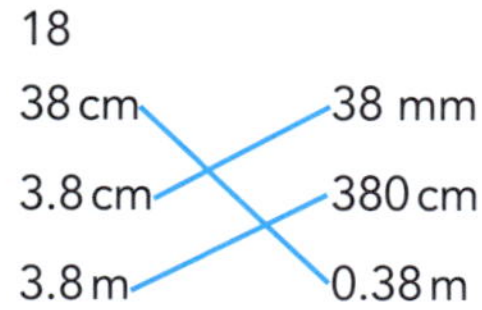

4 5,600

5 316

6 Award **2 marks** for the correct answer of 4.32

Award **1 mark** for evidence of the correct method with no more than one mathematical error, e.g.

7.2 cm × 60 = 432 cm

432 cm ÷ 100 = wrong answer

Mass (page 34)

1 4,310

2 Award **2 marks** for correct answer 1.625 kg or 1,625 g.

Award **1 mark** of correct reading of 875 g shown and evidence of a correct method,

e.g. 2,500 g – 875 g = error.

Do not award any marks for an answer of only 875 g with second step of problem omitted.

3 0.28

4 Award **2 marks** for correct answer 415

Award **1 mark** for a correct/appropriate method

e.g. 5,500 g – 520 g = 4,980 g
4,980 g ÷ 12 = error

or 5,500 g – 520 g = error

error ÷ 12 = error

Do not award any marks for evidence of only one step in the problem.

Capacity (page 35)

1 4,650

2 1,250

3 10

4 30

5 a) 10

b) 5

Money (page 36)

1 a) £26.25

b) £1.44

2 Award **2 marks** for correct answer hoody (£11.25) and cap (£6.99).

Award **1 mark** for evidence of total needing to be £18.24 but with error when adding two items.

Do not award any marks for only identifying £18.24 with no evidence of the second step.

3 Award **2 marks** for correct answer Kelsey £4.45 and Toby £3.20

Award **1 mark** for evidence of correct working but incorrect answers.

4 a) £18.75

b) £66.25

5 Award **2 marks** for correct answer £40.

Award **1 mark** for evidence of all steps of the problem.

Do not award any marks for an incomplete method.

Time (page 37)

1 Award **2 marks** for all correct.

Award **1 mark** for two correct.

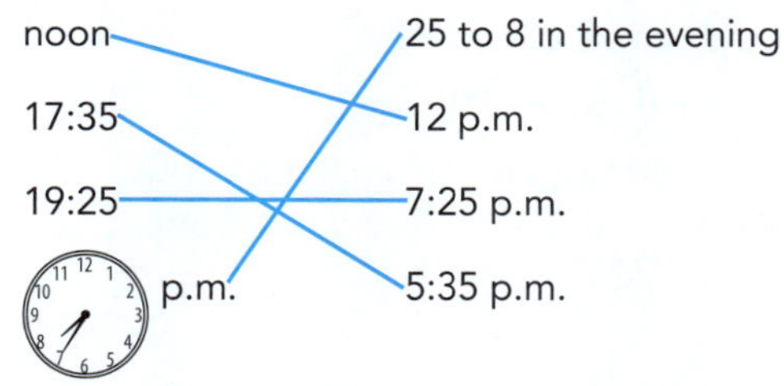

2 C (11:30 a.m.), D (12:00), B (21:35), A (10:40 p.m.)

3 8 o'clock or 8:00

4 11:45 a.m.

Time problems (page 38)

1 1 hour 45 minutes (accept 105 minutes)

2 13:50 or 1:50 p.m. (award the mark even if p.m. is not shown)

3 a) 10:30 or half past 10 b) £1.75

4 a) 1 hour 50 minutes (accept 110 minutes)

b) 51 minutes

c) 12:57 (accept 3 minutes to one)

Perimeter (page 39)

1 30

2 7.5

3 8

4 Award **2 marks** for correct answer 41.4 cm.

Award **1 mark** for 414 mm or for correct measurements of unlabelled sides identified but with error when calculating the perimeter.

Do not award any marks for correct measurements added to the diagram but with no signs of calculating the perimeter.

Estimating the area of irregular shapes (page 40)

1 Accept any shape with an area of $7\frac{1}{2}$ squares.

2 Shape B

3 Approximately $11\frac{1}{2}$ cm^2 (11.5 cm^2)
(accept an answer that is more than 11 cm^2 but less than $11\frac{3}{4}$ cm^2)

4 Award **1 mark** for an explanation that shows in some way that a better estimate or approximation would be 17–17.5 m^2. This may be through annotations on the diagram that clearly show that the area is more than 15 m^2.

Do not award any marks for an explanation that Asha has not counted the squares carefully.

Area by formula (page 41)

1 121 cm^2

2 A (36 m^2)

3 Any two sets of dimensions using factors of 36 (1 cm and 36 cm, 2 cm and 18 cm, 3 cm and 12 cm, 4 cm and 9 cm) or more sophisticated responses such as 5 cm and 7.2 cm

4 Award **2 marks** for correct answer 3 m^2.

Award **1 mark** for evidence of a correct method e.g.
4 m × 3 m = 12 m^2
3 m × 3 m = 9 m^2
12 m^2 – 9 m^2 = error, or for giving the answer in cm^2

Geometry – Properties of shapes

Drawing lines and angles (page 42)

1 Rectangle drawn accurately with 90° angles and sides of 6.5 cm and 3.5 cm

2 a) 7.5 (accept 7.3 to 7.7)

b) 35 and 55

(give margin error of 2° on either side of exact measure)

3 a) Angle a = 65°

Accept angles drawn in the range of 63° to 67° inclusive.

b) 104°

Accept answers in the range 102° to 106° inclusive.

2-D shapes (page 43)

1

Name of 2-D shape	Number of straight sides	Number of right angles
rectangle	4	4
right-angled triangle	3	1
regular hexagon	6	0
equilateral triangle	3	0

2

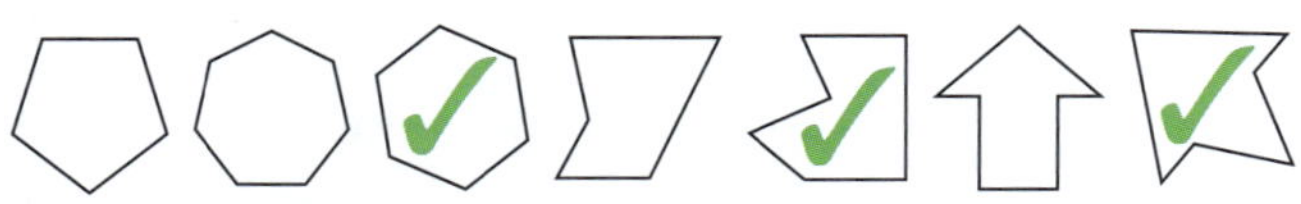

3 rectangle, rhombus

4

Name of shape	Number of pairs of parallel lines	Number of pairs of perpendicular lines	Total number of sides
irregular pentagon	1	2	5

3-D shapes (page 44)

1 5 faces, 9 edges, 6 vertices

2

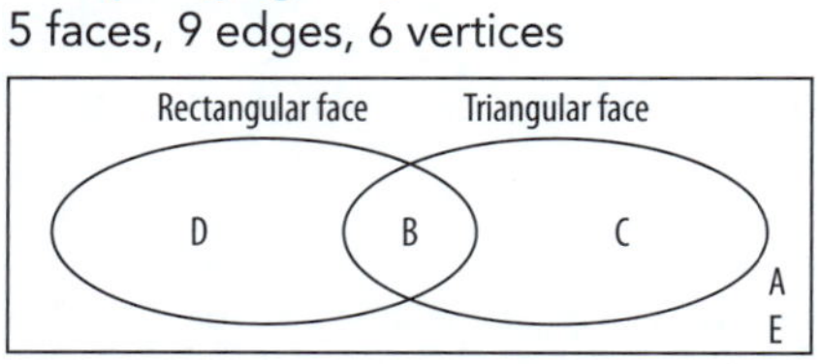

3 square-based pyramid

4 C and E ticked

5 A and C ticked

Angles and degrees (page 45)

1 120

2 50

3 a = 72 and b = 135

4 75

5 Award **2 marks** for correct answer 105.
Award **1 mark** for evidence of the correct method (i.e. 360 – 150 = 210, then divided by 2 = incorrect answer).

Do not award any marks for not recognising that angles at a point sum to 360° or for omitting a step in the problem.

Angles in triangles (page 46)

1 42

2 All angles labelled as 60° **or** sight of calculation 180° ÷ 3

3 a = 90° b = 55°

4 55° and 55°

5 a = 90°

b = 45°

Geometry – Position and direction

Coordinates (page 47)

1 **A** (2,2), **B** (6,4), **C** (3,6)

2 (8,5)

3 pentagon (accept irregular pentagon)

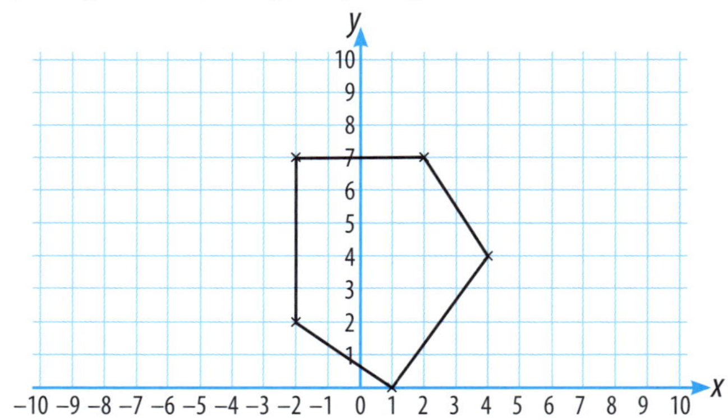

Award **2 marks** for a correctly named shape and correct shape drawn on grid as shown here.

Award **1 mark** for all coordinates correctly positioned but not joined or shape unnamed.

Do not award any marks if any coordinates are incorrectly positioned.

4 (5,7) and (8,7) **or** (5,1) and (8,1)

Translations (page 48)

1 (0,0)

2

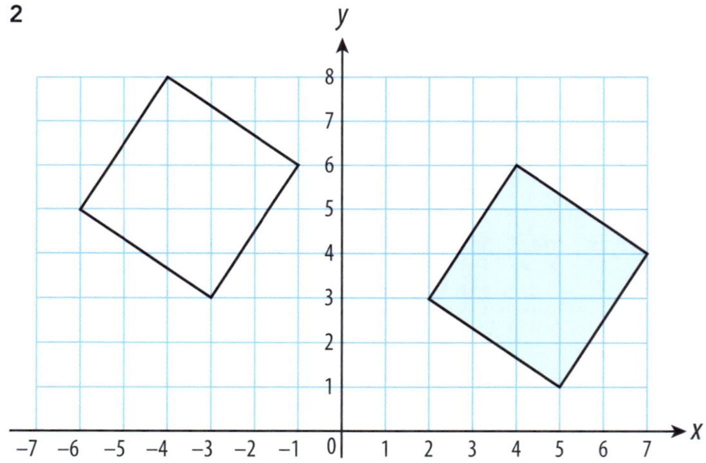

3 2 units/squares right and 3 units/squares down

Reflections (page 49)

1 C

2

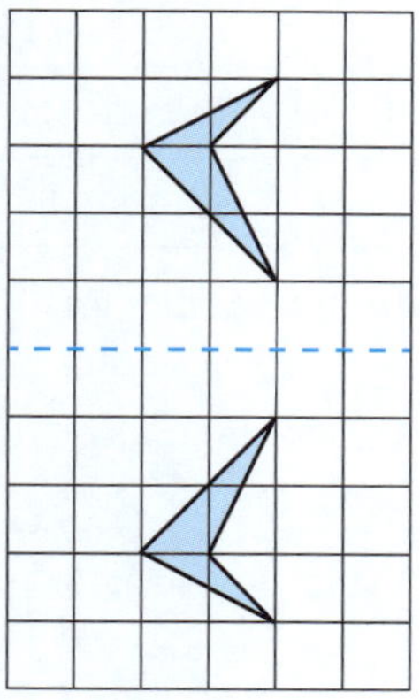

3

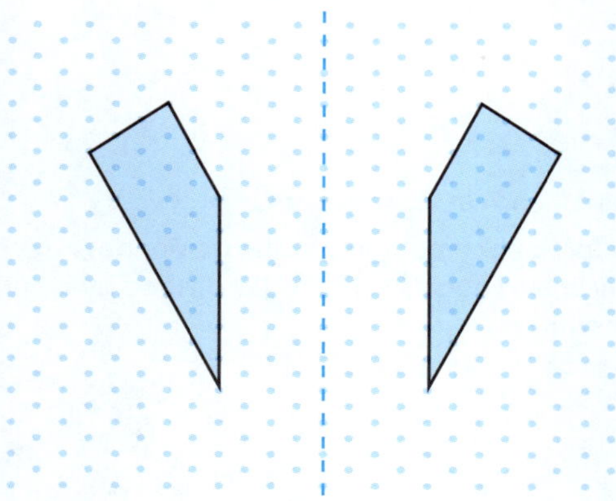

4

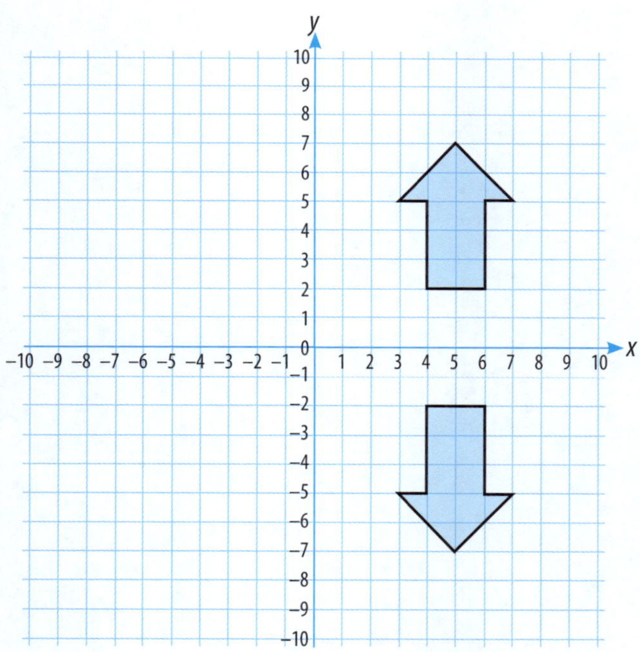

Reflective symmetry (page 50)

1

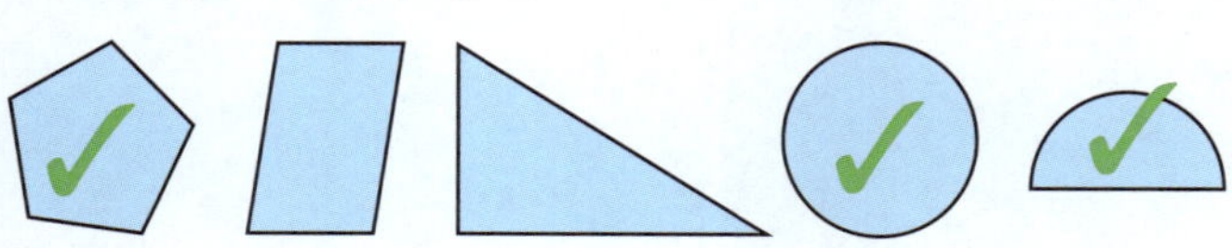

2

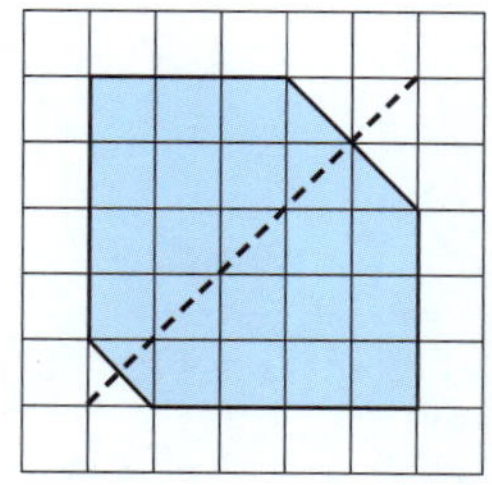

3

	Reflective symmetry	No reflective symmetry
Quadrilateral	E	C
Not quadrilateral	B	D A

4

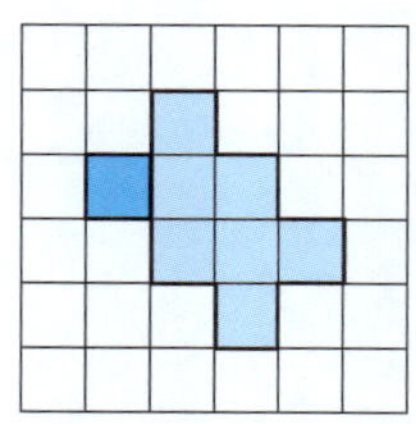

Statistics

Tables (page 51)

1

Favourite sport	Tally	Total
badminton	𝍸 𝍸 𝍸 𝍸 III	23
hockey	𝍸 𝍸 II	12
football	𝍸 𝍸 𝍸 III	18

2 a) £39.43

b) £3.74

3 a) Wednesday b) Kate and Jayden

4 a) $1\frac{1}{2}$

b) Award **2 marks** for correct answer $32\frac{1}{4}$.

Award **1 mark** for evidence of a correct method.

Do not award any marks for an incomplete method.

Pictograms (page 52)

1

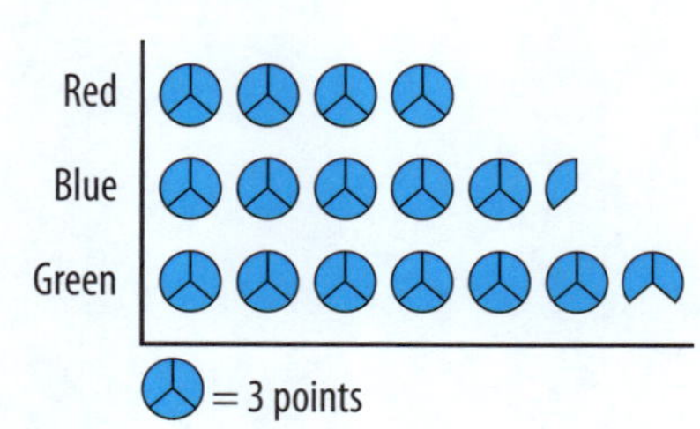

2 a) 10 b) 62 c) 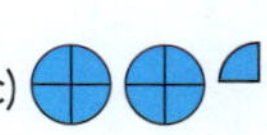

3 a) gardening and fashion
b) sports and games

Bar charts (page 53)

1 a) 13 b) 103

2 a)

	Object A	Object B	Object C	Object D
Mass (g)	350	500	550	650

b) Bar correctly drawn to show 450 g

3 a) 2

b) Accept a bar that is clearly more than £10 but less than £12.50

c) Award **2 marks** for the correct answer of 11.25

Award **1 mark** for evidence of the correct method with no more than one mathematical error, e.g.

12.50 + 25 + 17.50 + 22.50 + 11.25 = 88.75

£100 – £88.75 = wrong answer

Pie charts (page 54)

1 a) $\frac{3}{4}$ b) 90

c) YES, supported by an explanation that shows clearly that $\frac{1}{4} + \frac{1}{4} = \frac{1}{2}$

2 a) $\frac{3}{8}$ b) £30 c) £30

Line graphs (page 55)

1 a)

Day	4	8	14	20
Height (mm)	10	25	35	55

b) 67.5 (accept an answer that is greater than 66 but less than 69)

2 a) 5 b) 20 c) $2\frac{1}{2}$ hours

Mean (page 56)

1 9

2 Red = 6; Blue = 10; Yellow = 8

Total = 24 Mean is 24 ÷ 3 = 8

3 $2\frac{1}{2}$

4 10

5 57.5

6 64 minutes

Award **2 marks** for correct answer of 64 minutes.

Award **1 mark** for correct method of total of Monday, Tuesday, Wednesday, Thursday and Friday divided by 5 but an incorrect answer.